INTRABRANDING

The
Keystone
of
Corporate
Agility

Also by Marc Rudov

Brand Is Destiny: The Ultimate Bottom Line

*Be Unique or Be Ignored: The CEO's
Guide to Branding*

INTRABRANDING

The Keystone of Corporate Agility

MARC RUDOV

MHR Enterprises

MHR Enterprises

Intrabranding
The Keystone of Corporate Agility

By Marc H. Rudov

Published by:
MHR Enterprises
MarcRudov.com

ISBN: 978-0-9745017-7-2 (paperback)
ISBN: 978-0-9745017-8-9 (eBook)

Library of Congress Data
Rudov, Marc H.,
    Intrabranding: the keystone of corporate agility/Marc H. Rudov
    —1st ed.
    ISBN: 978-0-9745017-7-2 (paperback)
    1. Branding
    2. Marketing
    3. Sales
    4. Business strategy
    5. Management

Library of Congress Control Number: 2020906371

The difference between mere management and leadership is communication.

Winston Churchill
Former British Prime Minister

DEDICATION

To my late parents, Walt and Corinne, who rejoiced in my successes. They admonished me at an early age to speak clearly and persuasively, with good grammar. I'm forever grateful to you both, and I miss you.

Table of Contents

INTRODUCTION

If you've ordered your employees to wear facemasks, practice social-distancing, read the fraudulent *White Fragility*, and view America as systemically racist (a big, fat lie), and they rapidly obeyed you—*but still don't know your company's brand*—you've succeeded at spreading wokeness but failed at intrabranding: internally selling and enforcing your brand.

We live in tumultuous times. The COVID-19 pandemic has exposed the weaknesses of all companies, big and small. They are fighting for survival. The agile ones will emerge as stronger enterprises; the others will flounder or die.

Under former CEO Ginny Rometty (2012-2020), IBM struggled with its purpose and direction (brand), and revenue growth. Despite claiming IBM as not product-centric, it was. Embarrassingly so. Its ever-changing, tech-focused branding (messaging) was confusing, inept, and self-defeating.

On May 20, 2020, responding to COVID-19, IBM's new CEO, Arvind Krishna, announced a round of job cuts (the usual CEO move) *aimed at making the company more agile*:

"IBM's work in a highly competitive marketplace requires flexibility to constantly remix high-value

1

skills, and our workforce decisions are made in the long-term interests of our business."

Agility. Flexibility. Attributes of any enterprise that is beating its competitors by attracting and retaining customers, investors, and top talent. Hence, the subtitle of this book.

Watch the best football and basketball teams. What do they have in common? *Agility.* Skill isn't enough. Skill, poorly aligned and directed, begets chaos and failure.

Great teams, like great enterprises, are *agile*—they can perceive situations and proceed quickly, in real time. The roots of agility are *urgency, alignment, and communication.*

Leaders Communicate

When watching President Trump and his coronavirus taskforce deliver daily updates, in early March 2020, I heard differing and conflicting messages emanating from Trump, Vice President Mike Pence, Secretary of HHS Alex Azar, FDA head Stephen Hahn, CDC director Dr. Robert Redfield, response coordinator Dr. Deborah Birx, NAIAD director Dr. Anthony Fauci, and Surgeon General Dr. Jerome Adams.

It dawned on me that, while receiving such critical information daily was valuable, it was confusing and hard to remember. And, that the parties weren't on the same page compounded the difficulty. Moreover, there was no *central* place to read an organized, categorized accounting of these COVID-19 updates—and get survival advice. Unacceptable.

On March 9th, I initiated an intensive campaign to lobby the aforementioned parties with missives—via Twitter, emails, and high-placed mutual connections—admonishing them to create my desired central depot: *coronavirus.gov*.

My efforts paid off! That URL is now the official federal repository of COVID-19 info.

The clumsy government reaction to COVID-19 was my "aha moment," highlighting the effects of poor alignment and communication, thereby inspiring me to write this book.

ATTENTION
Vice President Pence
UPDATE AMERICANS QUICKLY
CREATE THIS INFO HUB
CoronaVirus.gov
(Recommended by MarcRudov.com)

Communications deficits beget failures in human relationships, business and personal, and in companies.

Can your employees, in *all* departments, recite your company's brand? If not, that's a huge problem for you.

In Chapter 7 of *Brand Is Destiny*, I introduced the "entroprise," a portmanteau of entropy (chaos) and enterprise, a corporation whose left and right hands have yet to meet. More likely than not, you work for or run an entroprise.

Uniformly conveying a corporate message (brand) *externally* will fail unless your *internal* ecosystem—C-suite, board, employees, PR firm, ad agency, law firms, CPA firm, marketing partners, distributors, and franchisees—know, grasp, support, and consistently utter said message.

The inability to do so is an impediment that plagues companies of all sizes, ages, industries, and geographies—and is why their branding activities fail or are suboptimal.

Namely, poor internal communication, both in-person and electronic, impairs corporate agility: the ability to think, decide, and act quickly.

After creating the corporate brand, the CEO must *sell it to the troops*, verify that they grasp and embrace it, *mandate* that they use it, and *prove* that they're complying.

I call this *internal* communication, education, selling, and enforcement process *intrabranding*.

Winston Churchill, British prime minister from 1940-1945 and 1951-1955, rallied his citizens and helped President Roosevelt defeat the Nazis in World War II. A masterful, oft-quoted speaker, he notably opined: "The difference between mere management and leadership is communication."

Poor communication, sadly, is the norm in most companies. C-suite executives easily and readily admit this.

Accordingly, effective enterprise leadership requires ubiquitous, impactful communication—to and from every employee, at every level—that translates into action.

Effective leadership also requires *followship*—massive employee cooperation—which only good leaders can engender.

Simply talking, sending memos, emailing, and handing down edicts through subordinates is *not* communicating, and certainly not leading. If a broadcast tower beams a signal all over the city, but few radios can receive it, or many receive it but few listeners heed it, communication doesn't happen.

INTRABRANDING TOPOLOGY

C-Suite
CEO

EFFECTIVE ENTERPRISE LEADERSHIP *REQUIRES* UBIQUITOUS, IMPACTFUL COMMUNICATION

Intrabranding.com
© 2020 Marc Rudov

Confusing communication can be worse than none. Dozens of survivors of the World Trade Center, during the 9/11 attacks, said that many who had begun to exit the south tower, following intercom instructions, returned to their desks after getting new intercom instructions to stay. Alas, about 600 people were trapped and died in the upper floors.

The Keystone

In architecture, the arch is one of the strongest structural elements. At its apex is a wedge-shaped keystone, depicted below and on the cover of this book, which holds the other stones in compression and in place, allowing them, as a solid unit, to bear tremendous weight.

The arch, therefore, is only as strong as its keystone.

THE ARCH & KEYSTONE

Intrabranding.com

As the middle colony of the original 13, Pennsylvania was and is called the Keystone State. It always has been at the center of America's economic, social, and political activity.

Accordingly, I liken a corporation, or any organization, to the arch and intrabranding to the keystone of that arch.

A brand, the customers' emotional connection to a vendor, determines that vendor's purpose and direction. Without successful *intrabranding*—the keystone—therefore, the enterprise will behave like an entroprise, bereft of the strength and agility to anticipate, perceive, and respond quickly to any situation. Consequently, branding—*without sound intrabranding*—will be a costly exercise in futility.

CHAPTER ONE

Branding Review

Knowing the basics cold is essential for success. So, I'm including in this and the next chapter an abridged and updated version of Branding Review and Market Review, respectively, from *Brand Is Destiny*.

Vince Lombardi, legendary coach of the NFL's Green Bay Packers during the '60s, was famous for harping on the basics. He gathered his players at the start of each season, football in hand, to review game fundamentals and reiterate team objectives: "Gentlemen, this is a football."

The CEO of a global American firm was convinced—and constantly reminded his staff, his employees, investors, equity analysts, and the media—that his company's name is the most-recognized brand in his industry. If you have the best brand, I asked him, why aren't you #1? He had never thought of that. I explained that a name is *not* a brand.

If a CEO doesn't know what a brand is, and doesn't know what his company's brand is, how can he sell it, externally or internally? He can't. Are you getting the picture?

Ergo, if you and your employees are unclear about the basics—*you wrongly think that a product or company name is a brand and a technology is a market*—then, absorbing and internalizing this book's principles and advice is essential to your success.

Sell Me This Pen

A good sales trainer often uses this powerful training technique: hand the "student" a pen, then ask him to sell it back. Typically, the student, without hesitation, launches into a pitch about the pen's attributes. Bad move. Student fails.

This technique made two appearances in the 2013 blockbuster, *The Wolf of Wall Street,* about Jordan Belfort, the sex-addicted, drug-addled, convicted penny-stock scammer, played by Leonardo DiCaprio.

After serving 22 months in prison and paying tens of millions of dollars in restitution, Belfort, a master salesman, became—*what else?*—a motivational speaker.

In one of the movie's final scenes, Belfort was the guest speaker at a sales seminar. He took the stage, scanned the audience, then stepped off the stage to approach several attendees in the front row. In succession, he handed each attendee his pen and asked him to sell it back. After hearing a few words *about the pen*, Belfort cut off the attendee to give it to the next one. *Nobody knew how to sell the pen.*

Why the failures? Most people believe that selling is about the *product*. Wrong. Most people also believe that branding is about the *product*. Wrong again.

Branding is about *customers*.

Early in the movie, before Belfort became rich, he was sitting with a bunch of his misfit accomplices, contemplating his future company. He asked the same question. Only one guy, Brad, took the pen and responded correctly: "Write your name on a napkin." Belfort: "I don't have a pen."

Bingo. Brad did what most salespeople don't: learn the customer's needs *before* discussing the product.

Note: People *don't* buy products. Why are you pitching them? People buy solutions and value, based on their needs.

Read the two paragraphs above, over and over. Repeat them to yourself when you go to sleep and when you awaken. This is the essence of branding. If you don't know how to sell, you can't brand—*because you don't understand customers.*

What a Brand Is and Isn't

A brand is a *customer-validated value proposition*. It is *intangible* but vital: it characterizes the *emotional connection* between customers and a vendor.

To be successful, your customers *must* have an *emotional connection* with your company, regardless of your industry. If that connection doesn't exist or is weak, your company is a generic, lookalike producer of commodities.

If you don't understand the emotional stress of putting one's career on the line by signing a corporate purchase order, you will fail at branding. Keep reading.

Branding isn't about awareness, as most people believe, as GEICO believes. It's about *connection*, as I've been asserting ad nauseam.

BRANDING IS ABOUT *CONNECTION* (not awareness) MarcRudov.com

One expresses the brand, verbally and in writing, in *customer language*. That means *no* internal or industry jargon, and *no* mention of your company and its products, services, and technologies.

Most critical: your company *must* walk its talk—*deliver the brand's promise.*

Because the brand establishes a company's purpose and direction, branding is the CEO's #1 priority: the cost of a deficient or nonexisting brand is a weak bottom line.

Contrary to convention, branding **precedes** products, customers, and revenues. The brand is the foundation of your company—so you must build it **first**.

Every time a customer, investor, or reporter asks you to repeat, once again, what your company does, you know that your brand has failed.

BRAND IS FOUNDATION
YOU *MUST* BUILD IT FIRST

BRAND IS FOUNDATION

MarcRudov.com

In addition, a brand is *not* a logo; a logo is *not* a brand. The logo, a graphical symbol, *represents* a brand but never *constitutes* one. So, if you just hired a design agency to create a new logo and label, that's all you have. *Never* announce that your old logo *was* your brand and new logo *is* your brand.

Moreover, many professionals in business and media circles blithely *but mistakenly* refer to company and product names, and labels and SKUs, as brands. **Unless customers are *emotionally connected* to a company name or product name, it's just a *word*—and that's true in most cases.**

What Is a Brand?

N⊘T a logo, label, or SKU
N⊘T a product name
N⊘T a product description
N⊘T a company name
N⊘T a company description

© 2017 MarcRudov.com

Example: Upscale/Downscale Restaurant

A restaurateur in a tony Silicon Valley town asked me to create a tagline for her up-and-coming bistro. She wanted to stand out on a street with many competitors.

You might get the wrong impression about branding, thinking it's exclusively for consumer companies. Actually, many people falsely believe that nonsense. In *Brand Is*

Destiny, I also used a consumer example—even though many of my clients are in the industrial and technology spaces.

My rationale is pedagogical: To broaden the appeal of this book and hasten the learning process, I'm using an example to which EVERYONE can relate.

I entered the eatery at dinner time, had the hostess seat me, ordered a meal, and began eating. Then, I circulated amongst the tables, clipboard in hand, intruding unexpectedly on the other diners. I sat down at their tables, while they were eating, to ask them three fundamental questions:

1. Why are you here?
2. Why do you keep coming back?
3. What do you like about this place?

The patrons typically were well-heeled, some famous in the area, *but dressed casually*—and they liked it that way. They gave me a page full of comments and feedback, which I took back to my office to distill into a few words.

Note: Writing lots of words is easy, but a waste of effort: *nobody will remember, repeat, or respond to them.* On the other hand, writing a few memorable, effective words—*the key to successful branding*—is difficult; few people can do it.

I synthesized the customer comments into a four-word tagline: *Casual Crowd, Classy Cuisine.* My client and her patrons loved it. They could identify with and relate to it.

That's the essence of branding: converting customers' language and emotions into a pithy, repeatable phrase and reflecting it back to them—a phrase that resonates with them and makes them feel *emotionally* connected your company.

When customers feel *emotionally* connected to your company, they deem it superior to your competitors (not a lookalike producer of commodities) and prove it with higher loyalty and more repeat purchases. In other words, the strongest brand (not product) wins. *Again, this axiom applies regardless of your industry or company size.*

To repeat what I wrote before: the words of your brand are meaningless if they emanate from an entroprise that can't execute and deliver on the promise of said words.

If your enterprise can't walk its talk, it will lack the credibility to succeed at intrabranding and, thus, branding.

GutShare™

After watching the stock indexes fluctuate wildly every day, because of the *emotions* surrounding COVID-19 stats, potential treatments and vaccines, and the selective reopening of America, I'm including this updated GutShare section.

GutShare is that portion of your customer's *gut* your company occupies. You *must* grasp the *impact of emotion on business behavior*, a phenomenon that many wrongly deny.

Note: People make *gut* (not cerebral) decisions and purchases. The gold (profit) is in the gut, the nexus of logic and emotion. All of your messaging *must* be aimed at customers' guts.

Stop making *cerebral*, factual pitches to customers and investors. Remember: *No emotion, no motion (purchase).*

The homepage reveals your company's DNA, its style of thinking, acting, communicating everywhere—internally, CEO speeches and interviews, PR, ads, sales, customer service, and brochures.

If your homepage reads like a bland, boring, fact-based nutrition label, bereft of emotion, customers will view your company, *if they view it*, as a murky, functional, operations-oriented outfit—badly in need of fixing.

Why do most people continue to sell cerebrally? First, they're not trained. Second, it's safe and uncontroversial, and nobody ever got fired for spewing generic jargon. Sadly, it's also utterly boring, unmemorable, and ineffective.

Only those who've never engaged in face-to-face selling and negotiating have trouble grasping GutShare. Worse, they don't comprehend the concept of corporate emotions.

There are three universal corporate emotions: ***power, reputation, and paycheck***. Your brand—and by extension your salespitch—*must* reflect these emotions.

Consider what's at stake when your customers put their asses on the line, in their companies, to buy from you: Will they gain or diminish their powers? Enhance or trash their reputations? Increase, shrink, or lose their paychecks?

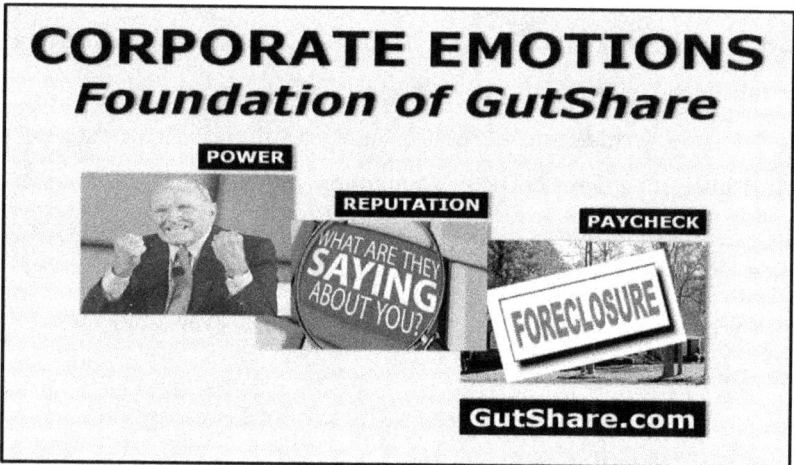

Brand Paves Product Path

Your company's brand sets its purpose and direction. It must, therefore, dictate which people you hire and fire, the processes you implement and retire, and the products you design, sell, and avoid. That's right: brand *before* products.

Most people laugh when I tell them this. But, not for long. Being a slave to products and technologies is a recipe for failure. The brand outranks you and everything you want to achieve, because it derives from the customer.

My metaphor for this paradigm is the org chart, seen below: *people, processes, and products report to the brand.*

Brand Outranks You

BRAND

PEOPLE PROCESSES PRODUCTS

© 2016 MarcRudov.com

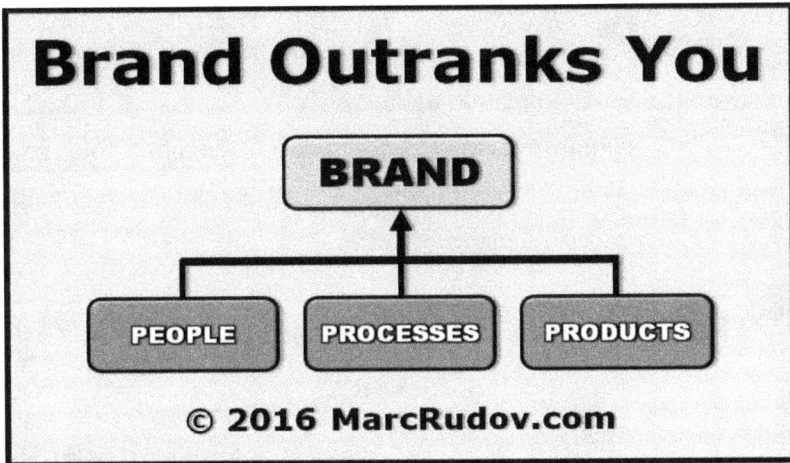

The head of HR for a major industrial corporation told me that he had hired a consulting firm to create an *internal* brand to unite the employees. The CEO approved this. My reaction: Why? This is a huge mistake (see the "Ditch Your Mission Statement" section below).

There's *one* brand, period, based on *customer needs*. Employees *must* learn and use that one.

Note: If you instruct your employees to learn more than one message, they'll ignore you and tune out. You will fail, your company will fail, and investors will bail.

This is why intrabranding is so critical. Again, if you falter internally, you'll do likewise externally—and vice versa.

I can't stress this point enough: Your company's brand is the road to its destiny. *It paves your product path.* That's why you must create it *before* developing any products. How else will you know which ones to build?

Merely extending previous products, because you've always done that, is shear laziness. Following the whims of your engineers is poor management.

In the figure below, products dot the path that your brand has paved. With a deficient brand, your products won't satisfy customers' needs and wants.

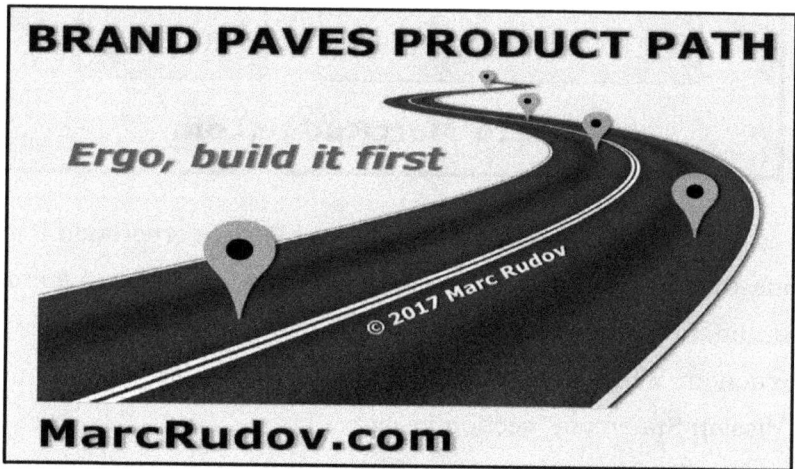

BRAND PAVES PRODUCT PATH

Ergo, build it first

© 2017 Marc Rudov

MarcRudov.com

Ignore Equity Analysts

It's bad enough that a plethora of public-company CEOs have trouble with branding and intrabranding, but they also succumb to messaging pressure from equity analysts.

CEOs must fight this pressure.

Equity analysts—who tend to be biased, myopic, and flamboyant—are extremely influential with investors.

The result is that these CEOs, fearful of criticism, frequently keep secret their future visions and often forgo executing the underlying strategies to convert those visions to reality.

This means that the analysts, not the CEOs, are calling the shots—and this is unacceptable.

Allowing outsiders to control your destiny is no way to run a company.

Ironically, this secrecy policy also keeps *customers* in the dark about your company's direction, negatively affecting the prices they'll pay for your products and services—thereby tanking your top and bottom lines.

According to researchers, "star analysts" (per tipranks.com)—but not "regular" analysts—not only affect stock prices but also the costs of and timetables for raising capital.

According to Boivie, Graffin, and Gentry, in a *Harvard Business Review* article from April 2016: *We found that a downgrade by a star analyst causes tremendous stock valuation changes that are not offset by the CEO having a good reputation. In other words, star analysts' reputations overwhelmed those of the CEOs they were covering in terms of shareholder reaction—even when star analysts downgraded firms run by star CEOs. Specifically, we found that a downgrade by star analysts increased the negative market reaction by 40%, regardless of a CEO's reputation. Thus, when a star analyst issues a downgrade, the CEO's reputation has almost no effect on the market reaction.*

In a jointly authored paper, in December 2017, economics professors Chen and Lu drew similar conclusions in China: *Sell-side financial analysts play a key role in collecting, interpreting, and disseminating company and market information to investors. Issuing "buy" and "sell" recommendations is an important part of an analyst's job and one of the most visible ways for them to express their opinions on the stocks and markets they cover. In an information market such as the one of financial analysts, since the product is ex-ante hard to evaluate, investors may reply on outside certification, such as award-wining status of an analyst, to infer the quality of his or her recommendations. In line with this argument, a large body of literature in finance and accounting have documented that investors react abnormally more to stock recommendations by award-winning financial analysts (hereafter "star analysts") than those by other analysts.*

Canadian economist François Derrien asserted in 2012: *Not only do analysts influence the price of shares via their investment recommendations and reports, the information they provide also directly affects the cost of the capital of the firms they cover and, more indirectly, their financing and investment decisions.*

Dos Doszhan, CEO of Stockmetrix, averred in May 2018: *Active trading based on the Wall Street analysts' recommendations regarding the S&P 500 companies turned out to be more profitable than [Warren] Buffet's passive investing approach. It seems like you can trust the analysts after all.*

In essence, the aforementioned paints a picture of CEOs passively dancing to the tunes of equity analysts, who have gained too much power—power devoid of checks and balances! This unfortunate situation has occurred over time, is counterproductive, and must end.

Surrendering to star analysts is not a solution for controlling your company's brand and, therefore, its destiny.

I recommend taking a stand. Emulate Donald Trump and Elon Musk. Be clever. Circumvent the analysts, *legally and wisely*, by communicating directly with investors and customers—frequently and compellingly.

Twofold objective: demonstrate your company's unique value and put the analysts on defense.

Build a TV studio at your headquarters. Make noteworthy status videos for your website. Provide live and taped video tours of your plants. Conduct interviews with happy customers. Use your imagination and creativity.

In other words, change the game.

Ideal scenario: Analysts are still involved but no longer in control. You are in control.

Or, you can continue your private griping about life as a marionette.

Ditch Your Mission Statement

It seems that every company, no matter its size or industry, feels compelled to construct a mission statement—but rarely obligated to create a brand.

Some also craft a "vision" statement. More nonsense.

Many believe the mission statement shapes company direction and rallies the troops. It does not.

Typical creation process?

Form a buzzword-happy, politically correct committee. Google boilerplate examples. Attend seminars. Read white papers. Hire a consultant. Emulate competitors' nebulous blatherings.

The result? Useless, generic, internally oriented, forgettable, uninspiring gobbledygook.

Want to know whether employees have any clue about their company's purpose and direction—what the brand defines—and whether they're rowing in the right direction, the same direction? Make a purchase. Then, call customer service when the process inevitably goes awry.

What you'll discover, after encountering a problem, is that the vendor's left and right hands have yet to meet—certifying a lack of corporatewide buy-in of purpose and direction. Yet, this supplier has a mission statement and maybe a vision statement.

Let's examine the mission statements of two of the top-three companies on the 2020 *Fortune 500* list: Walmart and ExxonMobil.

Walmart's mission statement: "We help families save money so they can live better." Walmart's brand, expressed in its tagline: Save Money. Live Better.

A rare example of mission and brand being identical. And, it's customer-focused, concise, memorable, meaningful, and repeatable. It reflects what Walmart's customers desire, in their language.

A more-common example is ExxonMobil's "Who We Are" statement: "ExxonMobil, one of the world's largest publicly traded energy providers and chemical manufacturers, develops and applies next-generation technologies to help safely and responsibly meet the world's growing needs for energy and high-quality chemical products."

Huh? Functional and operational jargon. Nothing about customers. I guarantee that no ExxonMobil employee can remember this committee-derived, meaningless, generic word salad, and won't even try. ExxonMobil's brand? No clue.

The only way to rally your employees to work as a team, for a single purpose, is to create and articulate a *brand*, as Walmart did. A brand is a concise, memorable, repeatable, *customer-validated* value proposition, expressed in *customer* language.

WARNING: If you ask your employees to memorize and internalize multiple statements (brand, mission, vision), they'll ignore all of them and work according to their own agendas.

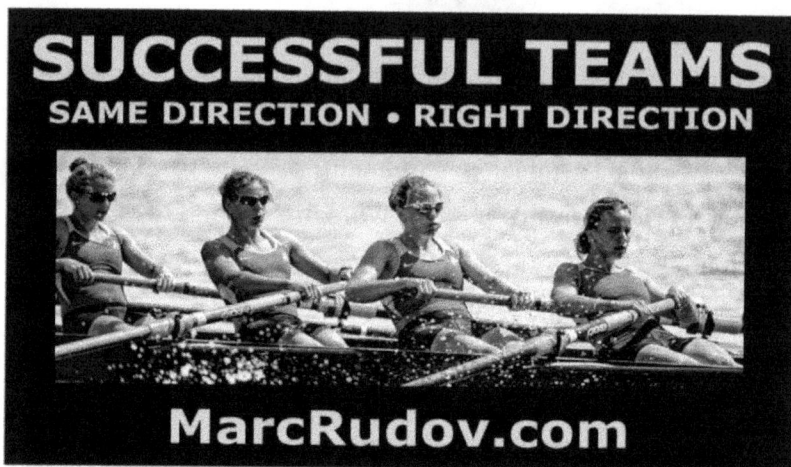

As I've written ad nauseam, a compelling brand drives every activity at your company: the people you hire, the processes you implement, and the products you build.

A strong, effective brand unifies the enterprise. It puts all employees, in all departments, on the same page. It distills all corporate communications into a single message, or pitch, for all constituents—ending the usual internal and external confusion and conflicts.

Ditch your mission statement and your vision statement, if you have one. Each is a useless corporate cliché that nobody will recite, remember, or adopt.

Multiple messages disorient your employees, confuse investors and reporters, and muddle your true mission: to magnetize your customers.

Your employees must row in the same and right direction. Otherwise, branding and intrabranding will fail.

CHAPTER TWO

Market Review

Many words in the business lexicon are misused and incorrectly interchanged, but none more so than *market*. I covered this in my prior two books, but review is paramount.

Let's look at fishing. The root of fishing is *fish*. Nobody on this planet would argue with that.

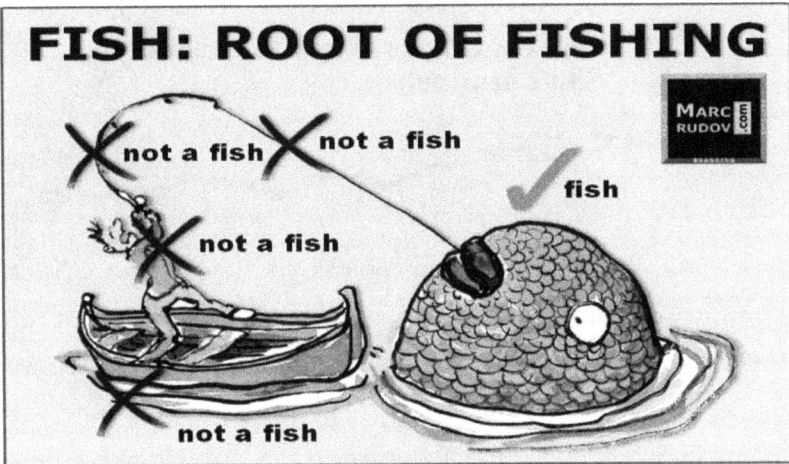

FISH: ROOT OF FISHING

The *line* is not a fish. The *pole* is not a fish. The *boat* is not a fish. The *fisherman* is not a fish. *Only* the fish is a fish.

Regardless of country or language, this universal acceptance prevents misunderstandings. Seems simple, doesn't it?

Market also has only one meaning: *customers*, current and potential. Get that wrong and you'll jeopardize market*ing* and branding. Kind of important, right?

Yet, corporations, consultants, TV commentators, and business publications misuse market every day. I could fill an entire book with examples. Here's a recent one from NatGeo.

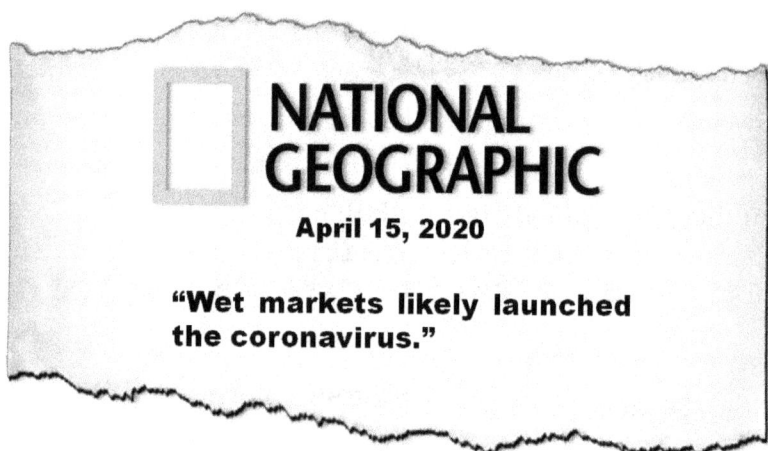

NATIONAL GEOGRAPHIC

April 15, 2020

"Wet markets likely launched the coronavirus."

All the evidence now points to the Wuhan (China) Institute of Virology as the source of COVID-19, but the reports for weeks had been blaming Wuhan's so-called "wet markets," where customers buy fresh meat, fish, produce, and other perishable goods—as opposed to durable goods like fabric and electronics.

These exchanges *should* be called wet market*places*, not markets, but most people don't know the difference.

Market *does not* equal product, service, technology, industry, or place to purchase same—even though people mistakenly do so, everywhere, every day.

Market, the *demand* (customer) side of commerce, is people. Describe them by *their* characteristics, not yours.

Industry is the *supply* side. Market*place* is the nexus of the two. Confuse any of this, and you'll myopically focus on your products, instead of your customers. Such is the annoying custom in high-tech and industrial precincts. This is *producting,* not marketing.

Market Research

Concomitant with our review of market is a revisit of market research, which means studying *people.*

A common misconception is that market research is about customers' views and uses of *products*. Wrong. And, why wouldn't that misconception be the case, given how many people falsely believe a product is a market?

Market research comprises studying how *people* live and work, based on *their* characteristics and demographics, without the interjection of your industry, your company and its products or services, and your technologies.

By understanding the wishes, needs, and problems of *people* in all walks of life, in all occupations, clever enterprises then can devise the appropriate solutions.

Asking people to imagine products that will solve their problems or grant their wishes—*which too many in the business world believe is market research*—is futile. The job of conceiving products is *yours*, not that of your customers.

As I've stressed numerous times and in numerous places, understanding customers comes from *knowing* them, not reading their social-media posts. Create as many opportunities as possible to meet them, face-to-face.

Analyzing customers' perceptions, impressions, usage, praises, and criticisms of your products, once you conceive or build them, is called *product* research. Conducting product research, *without having done market research,* is a total waste of time.

So, never confuse market and product research, both of which your company must continuously undertake and perfect.

Suing the Plaintiff?

A useful way to illustrate the importance of properly using business terminology is to examine what happens in other fields.

Imagine legal professionals (lawyers, judges, bailiffs, and court clerks) interchanging the following terms: plaintiff vs. defendant; petitioner vs. defendant; and appellant vs. appellee. The first parties bring actions, the second defend or respond to them.

What would happen if the US District Court in the Southern District of New York, in January 2020, allowed Tulsi Gabbard's lawyer to file her defamation lawsuit against Hillary Clinton (which Gabbard dropped in May 2020) as depicted below? Total chaos. Judicial meltdown.

UNITED STATES DISTRICT COURT
SOUTHERN DISTRICT OF NEW YORK

Tulsi Gabbard and Tulsi Now, Inc.,	Civil Action No. 20-cv-558
Plaintiffs, **(or Defendants)**	**COMPLAINT**
v.	**JURY TRIAL DEMANDED**
Hillary Rodham Clinton,	
Defendant. **(or Plaintiff)**	

Perhaps you scoff at my fictitious analogy as silliness. Why, then, do you blithely say wireless *market*, which doesn't exist? Getting the picture? FYI, it's a wireless *industry*.

People say they buy food at the market, trade stocks on the market, put their houses on the market, and compete in the job market. These are **marketplaces**. Using the wrong terminology causes ambiguity, which begets failure.

Unify Your Pitch

The epigraph of this book is from Winston Churchill, famous communicator and prime minister of Great Britain from 1940-45 and 1951-55: "The difference between mere management and leadership is communication."

Communication and branding are synonymous.

In April 2020, the *Pittsburgh Business Times* asked a dozen local CEOs how they're navigating the uncertainty of the COVID-19 pandemic. Nish Vartanian, CEO of MSA Safety Inc., a global designer and manufacturer of safety products, like PPE, for the oil and gas, fire, and construction industries, responded this way:

"Communication. Communication. Communication. We have had over 90 different communications that have gone out to our workforce around the world, tied to COVID-19, and we continue to produce a steady stream of communication to both internal and external stakeholders. We were lucky to a degree since we manufacture globally, and we have a facility in China. We started to assess the situation very early, and we put our crisis-management team together on Feb. 6th. We were able to learn what our Chinese organizations did."

An enterprise *cannot* succeed unless every employee, at every level, in every department is on the same branding page, speaking the same language to all internal and external constituents. If you can command every employee to social-distance and wear a mask, you can do likewise with your company's brand. The figure below depicts this axiom.

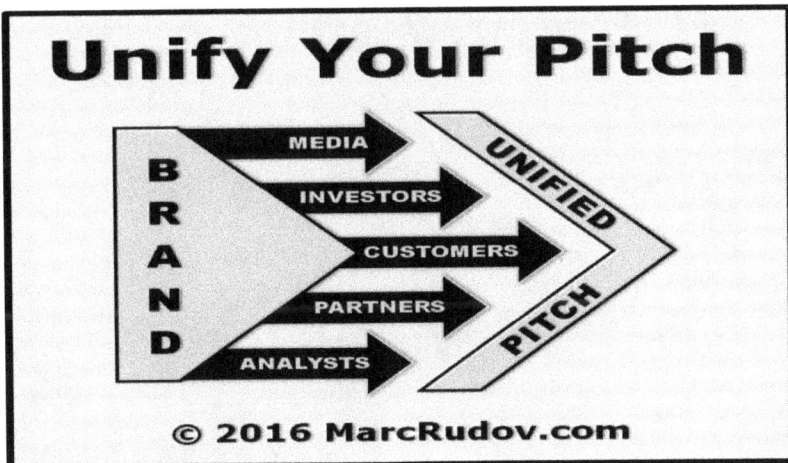

Unify Your Pitch

BRAND

MEDIA
INVESTORS
CUSTOMERS
PARTNERS
ANALYSTS

UNIFIED PITCH

© 2016 MarcRudov.com

Bottom Line

To excel at marketing and branding, you must know the definition of market and ensure that your employees and advertising and PR contractors do likewise.

Imagine the skipper and crew of a yacht confusing starboard with port. Might there be chaos? I think so.

Successful intrabranding requires, throughout your enterprise ecosystem, knowing and communicating the proper marketing and branding terminology.

CHAPTER THREE
Leadership & Intrabranding

Failure in intrabranding is the CEO's fault, regardless of his length of tenure. The CEO's leadership style, including taking a halfhearted or dim view of branding (which many do), can make or break any company.

The CEO has a choice: perpetuate or destroy inertia and tradition. Focus *outside* the building, not inside.

In the C-suite of one Fortune 500 company I advised is a hallway lined with portraits of past CEOs. Tradition and history have long dominated this company's culture. Those portraits *should be* of customers enjoying the firm's products.

What Is Leadership?

Discussing leadership makes no sense if we don't define it, just as we defined brand, branding, and market.

A leader is a person who easily gets others to follow. Without leadership *and* followship, intrabranding can't work.

33

In an August 2016 article in *Military Review*, Captain Charles R. Gallagher, from the Center for Army Leadership at Fort Leavenworth, opines wisely about the difference between management and leadership:

> "...leadership is an actionable behavior that can be learned. Leadership is not based on a person being in a traditional position of power or authority. Leaders are not necessarily generals, commanders, platoon leaders, or team leaders. Though the preceding groups often provide leadership, the definition also includes the emergent leader such as the rifleman who in a desperate situation leads his fellow soldiers through a firefight due to the absence of those in assigned leadership positions. Despite this soldier's lack of formal authority, he is still able to provide *purpose, direction,* and *motivation* through his behavior. John Kotter argues that the leadership process consists of establishing direction, *aligning* people, motivating, and inspiring. These actions are consistently different from management." [Emphasis added.]

Notice that Gallagher uses four leadership concepts critical to intrabranding: purpose, direction, motivation, and alignment. He explains later in his column that management is the nuts-and-bolts planning, organization, and execution of the business. One can be a good manager and a terrible leader. A great leader, though, *must* be a good manager: In the end, people will stop following someone who can't execute.

Does Military Background Help?

What kind of background best prepares a CEO to be a leader? I guarantee that universities do *not* provide leadership training—unless you consider leadership wanting to eliminate the First Amendment, protesting at the drop of a hat, or being perpetually aggrieved.

Is leadership learned through trial by fire? Copying executives in the workplace? How about military experience? At one time, almost all American men served in the armed forces. Today, not so much.

In January 2014, Carola Frydman and Efraim Benmelech, associate professors from Boston University and Northwestern University, respectively, reported their findings on this topic to the National Bureau of Economic Research. According to their research, the percentage of corporate CEOs with military experience steadily dropped from 59 percent in 1980 to only 6.2 percent in 2014.

Does this matter?

Frydman claimed that, historically, economists would consider the most effective CEO the one who maximized value by implementing the firm's policies.

Benmelech, who was born, raised, and served in the military in Israel, was especially interested to learn the military's impact on corporate life.

The two researchers found that military CEOs tend to make ethical, conservative decisions—and to be particularly adept at leading firms under duress.

For two years, they gathered biographical data on CEOs from the 800 largest US firms, from 1980 and 1991, and from approximately 1500 publicly traded US firms from 1992 to 2006. They tracked whether the chief execs served in the military—if so, in which branches, at what ranks, and the durations of service.

They found that military CEOs:

1. Perform better, and better under pressure, in industries going through decline or in distress.
2. Are typically prepared to make tough decisions and show leadership in tough times.
3. Are normally conservative and risk-averse, less likely to make bold investments in physical capital or R&D.
4. Are up to 70 percent less likely to engage in corporate fraud.

It makes sense that military CEOs, on average, are better in crises and situations of grave distress than their civilian counterparts. Usually, those are the times when boldness and risk are counterproductive. But, not always.

General (and eventual president) Ulysses S. Grant won the Civil War—a theater of grave crises and distress—*because of his boldness, creativity, grasp of details, and neverending risk-taking.* Moreover, he spent significant time at battlefronts witnessing firsthand what was happening. Today, he is viewed in military circles as a model of warfare brilliance on all three levels of leadership: strategic, tactical, and operational.

We know nothing about the other CEOs in the Frydman-Benmelech study, nor do we know the conditions of their companies. Remember the "conglomerate era" of the '60s, when companies idiotically bought completely unrelated businesses? It took decades to deconglomerate them.

Many CEOs are incompetent, unimaginative cowboys and sloths, who make illogical, ego-based decisions—or make none at all, allowing their firms to languish. Perhaps, in comparison, steady, by-the-book military CEOs looked stellar. We don't know.

Culture

In a July 2016 study, published in *The Journal of Applied Psychology*, researchers at Georgia State University, Arizona State University, the University of South Australia, and Auckland University of Technology concluded that, to improve enterprise performance, the CEO's leadership style should *differ* from its culture.

Based on data collected from 119 CEOs and 337 top management team members, in 119 American hardware and software companies, the researchers found that CEOs who adopt a leadership style similar to the organization's existing culture have a *negative* impact on a firm's performance—a discovery that contradicts conventional beliefs.

"Consistencies between CEO leadership and culture create redundancies," said Chad Hartnell, assistant professor in the J. Mack Robinson College of Business at Georgia State. "Leaders who are culture conformists are thus ineffective. CEOs who lead in a manner different from the culture benefit companies because they provide resources to the organization that the culture does not."

According to the researchers, organizational culture comprises shared values and norms that influence employee behavior. They claim that, in a **task-based** [branding-oriented] culture, employees focus *externally* on anticipating customers' needs and preferences and monitoring competitors' behaviors.

But, in a **relationship-based** culture, employees focus *internally* on issues such as coordination, participation, and communication.

"Similarities between leadership and culture can produce a myopic focus on things that have worked in the past while precluding employees from acquiring other resources or processes that could enhance success," Georgia State's Hartnell said. "CEOs should be mindful about focusing employees on important outcomes and processes that cultural signals may overlook."

Hartnell continued, "Not all differences between leadership and culture are positive. If a leader's approach is oppositional or confrontational, he or she will likely be met with resistance and resentment. A leader who challenges or discards every assumption about what has worked in the past creates uncertainty, ambiguity, and skepticism among the organization's employees.

"Leaders must search diligently for what isn't currently being handled by the culture and fill in the gap. They should adopt a leadership style that builds upon the positive aspects of the existing culture, contributing to the culture without undermining it."

Fake Leadership

During a March 2019 installment of the Goldman Sachs interview series, "Talks at GS," Rosalind Brewer, COO of Starbucks, uttered: "One of the things I realized is that you have to manage people the way they want to be managed."

Is *that* management? No. It's fake leadership, a lie.

Walk into any highly regimented Starbucks, and you'll see the burned-out human robots at work.

Why did Brewer make such a ludicrous statement?

Simple: Corporate virtue-signaling, which has become the unfortunate norm these days.

In August 2019, I described this fake leadership in a piece about 181 woke CEOs of the Business Roundtable. These virtue-signaling titans professed a "new purpose" of the corporation: less emphasis on profits and shareholder value, more devotion to "community affairs" (translation: controlling guns and fighting climate change). Their collective goals:

- Delivering value to our customers
- Investing in our employees
- Dealing fairly and ethically with our suppliers
- Supporting the communities in which we work
- Generating long-term value for shareholders.

Today's CEOs crave to be "woke" to whining, holier-than-thou Millennials and Gen-Zers, and the growing swarm of anti-capitalists in America. Profit is now a dirty word.

Apparently, unless today's corporations toe the politically correct line, they have trouble recruiting and retaining history's first virtuous employees.

Jamie Dimon, chairman and CEO of JP Morgan Chase and a Roundtable signatory, asserted this:

> "Major employers are investing in their workers and communities, because they know it is the only way to be successful over the long term. These modernized principles reflect the business community's unwavering commitment to continue to push for an economy that serves all Americans."

The irony: everything listed in the five goals above is *required to generate long-term value for shareholders.* Nothing new here. These chief executives just discovered branding.

The essence of branding is understanding customer needs, communicating those needs back to the customers—in their language—and delivering unique value to them.

No company can succeed, on a sustainable basis, by mistreating employees and suppliers. Nor can it succeed by letting green, infantile, outraged employees call the shots.

Alas, many CEOs ignore or don't grasp this concept.

America's top leaders announcing a "modern standard for corporate responsibility" is akin to a car-owner bragging about regularly checking his tire and oil pressure.

One line-item above, though, bears closer inspection: *supporting the communities in which we work.*

What does that mean? Buying shirts for the local Little League teams? Becoming a substitute governmental agency, doling out largesse to keep the politicians away? We don't know. But, companies can't do that: ***they must earn profits***.

If companies don't maximize shareholder value, they'll lose shareholders, without whom they cannot function.

Trying to sustain long-term value with a short-term, shortcut mentality doesn't work. Being woke is a joke.

Contrast this with Elliott Management, an investment firm, which, in September 2019, owned roughly $3 billion (1%) of AT&T stock. Elliott wanted AT&T management to shape up or ship out, including dumping acquisitions DirecTV and TimeWarner. Elliott's demands were about *profits, profits, profits*—not gun control and fighting climate change.

The Do-Gooder B Corporation

The July 7, 2020, issue of *Fortune* featured a piece on food giant Danone's quest to become certified as a B Corp (B Corporation) by 2025.

Wokeness has no end. What does this entail? Danone will be held to strict standards of "verified social and environmental performance, public transparency, and legal accountability to balance profit and purpose."

Expressed on Danone's website:

Our ambition to obtain this certification is an expression of our long-time commitment to sustainable business and to Danone's dual project of economic success and social progress.

It is a significant step toward making sustainable business mainstream—and which we believe to be the future.

Seeing a trend here? CEOs, to mollify Millennials, Gen-Zers, academia, and the progressive media, have become social-justice warriors. This virtue-signaling is a lame, phony substitute for branding and will backfire.

Per the B Corporation website: As B Corporations and leaders of this emerging economy, we believe that:

- We must be the change we seek in the world
- All business ought to be conducted as if people and place mattered
- Through their products, practices, and profits, businesses should aspire to do no harm and benefit all
- We act with the understanding that we are each dependent upon another and thus responsible for each other and future generations.

Certified B Corporations are businesses that meet the highest standards of verified social and environmental performance, public transparency, and legal accountability to balance profit and purpose. B Corps are accelerating a global culture shift to redefine success in business and build a more inclusive and sustainable economy.

Society's most challenging problems cannot be solved by government and nonprofits alone. The B Corp community works toward reduced inequality, lower levels of poverty, a healthier environment, stronger communities, and the creation of more high-quality jobs with dignity and purpose. By harnessing the power of business, B Corps use profits and growth as a means to a greater end: positive impact for their employees, communities, and the environment.

Who does the certifying? Standards analysts at the non-profit B Lab's Pennsylvania, New York, and Amsterdam offices. The standards for B Corp Certification are overseen by B Lab's independent Standards Advisory Council. They measure how a company's operations and business model impact its workers, community, environment, and customers—from its supply chain and input materials to its charitable giving and employee benefits. Corporations must sign a Declaration of Interdependence and pay B Corp an annual fee for the certification.

Note: B Corp-certified companies have, in essence, changed their governance structures, reporting to *two* boards of directors: one representing the shareholders *and* one representing the woke police. This insane political correctness will kill their brands.

Don't exhale yet. There's another woke entity called the *benefit corporation*, often confused with the B Corp. It's a traditional corporation, not certified or audited like a B Corp, obligated to higher standards of *purpose*, accountability, and transparency. Wokeness has no bounds.

In Chapter One (Branding Review), I counseled you *not* to create a mission statement: the *only* messaging employees should memorize is your brand, *which sets your company's purpose and direction.* Alas, B Corps are brainwashing their employees with social-justice nonsense, like climate change, that has *nothing* to do with satisfying customers. Guess what B Corp employees *won't* be memorizing: the brand.

Guts at Goya

On July 9, 2020, Bob Unanue, CEO of Goya Foods, was a guest of President Trump in the White House. He praised Mr. Trump, "We're all truly blessed at the same time to have a leader like President Trump, who is a builder." Seems OK, right? Wrong. It's the cancel-culture era.

Progressive Latinos (including AOC and Lin-Manuel Miranda) went berserk, of course, calling for everyone to boycott the $1.5-billion/year, Secaucus, NJ, seller of beans and rice for the Spanish and Latin markets.

The blowback against the woke boycotters—the "buy-cott"—was swift. Francisco Marte, secretary-treasurer of the Bodega and Small Business Association, was angry at the mob for "enforcing political conformity on one of this country's most successful job creators" while "harming Hispanicimmigrant-run stores that work long hours to make ends meet amid a challenging economic and health crisis."

The next day, Unanue, who once attended a Hispanic Heritage Month event at Obama's White House, said on *Fox & Friends*, "We were part of a commission called the White House Hispanic Prosperity Initiative, and they called on us to be there to see how we could help opportunities within the economic and educational realm for prosperity among Hispanics and among the United States."

Mr. Unanue called the pushback against him visiting the White House and praising Trump "suppression of speech." "So, you're allowed to talk good or to praise one president [Obama], but you're not allowed to aid in economic and educational prosperity [with Trump]? And you make a positive [Trump] comment, and, all of a sudden, it is not acceptable?"

Unlike most American CEOs, who tend to be spineless, woke, virtue-signaling wimps, Unanue made it clear that he's not apologizing for his remarks supporting Trump's economic policy and would not turn down other future invitations.

Imagine how peaceful America would be if other CEOs would emulate Unanue's leadership and tough stance against the mob!

Alas, as more CEOs supinely enable the mob, it grows in audacity and destructiveness. Alternatively, because gutsy Mr. Unanue bucked the mob, stores carrying his products were sold out of them almost immediately. Got it?

If you can't take a stand, you don't have a brand.

Standing Up to BLM

Since the murder of George Floyd in Minneapolis on May 25, 2020, Black Lives Matter (BLM) has dominated the news. According to Rasmussen, BLM is more popular than Trump—despite its heritage of Marxism, extreme violence, anti-Semitism, and intent to destroy the nuclear family.

Said BLM cofounder Patrisse Cullors, "We are trained Marxists." Warned Hawk Newsome, president of Greater New York BLM, "If this country doesn't give us what we want, then we will burn down this system and replace it. " Got it?

After BLM torched her house, Olympia, Washington's mayor Cheryl Selby, who initially praised the group, labeled its protests "domestic terrorism."

Leo Terrell, a black civil-rights lawyer in Los Angeles, opined in *Newsweek* on July 3, 2020:

> "Black Lives Matter is comprised of modern-day charlatans who learned the profitable way to protest from the Al Sharptons of the world. Style over substance. Chaos over real change."

Jason Whitlock, a black sports analyst, opined thusly, on July 6, 2020, to Tucker Carlson of Fox News:

> "By adopting the changes demanded by Black Lives Matter, the National Football League will never be seen the same again because of the failure of leadership."

"If the NFL starts out its season with everyone standing for 'Lift Every Voice and Sing,' the black national anthem, and then virtually everyone on the field taking a knee when 'The Star-Spangled Banner' plays, I think it's going to be—if you remember the show *Happy Days*—the jump-the-shark moment when it's like, OK, *Happy Days* is over."

No matter. Woke CEOs have rushed to apologize for **nonexistent systemic racism in America**, glamorize BLM, *and send BLM hundreds of millions of dollars*—in other words, funding Marxists out to overthrow America.

Apparently, enabling Marxist BLM with guilt-cash and ransom, that shareholders likely did not approve, is justified recompense for exporting inner-city jobs to China. Right?

The mobs, inside and outside companies, are now controlling CEOs: *the antithesis of leadership.*

CHAPTER FOUR

Enterprise Governance

The East India Company

To understand why corporations typically follow a certain pattern of management, let's explore how that pattern originated and developed.

The first megacorporation was the East India Company (EIC). The "Boston Tea Party" was in protest of the British Parliament's tyrannical Tea Act of 1773, designed to save the faltering EIC by lowering its tea tax and granting it a virtual monopoly on the American tea trade. This lower tax allowed EIC to undercut tea smuggled into America by Dutch traders.

National Geographic described EIC in September 2019:

Think Google or Apple are [sic] powerful? Then you've never heard of the East India Company, a profit-making enterprise so mighty, it once ruled nearly all of the Indian subcontinent. Between 1600 and 1874, it built the most powerful corporation the world had ever known, complete with its own army, its own territory, and a near-total hold on trade of a product now seen as quintessentially British: Tea.

At the dawn of the 17th century, the Indian subcontinent was known as the "East Indies," and—as home to spices, fabrics, and luxury goods prized by wealthy Europeans—was seen as a land of seemingly endless potential. Due to their seafaring prowess, Spain and Portugal held a monopoly on trade in the Far East. But Britain wanted in, and when it seized the ships of the defeated Spanish Armada in 1588, it paved the way for the monarchy to become a serious naval power.

In 1600, a group of English businessmen asked Elizabeth I for a royal charter that would let them voyage to the East Indies on behalf of the crown in exchange for a monopoly on trade. The merchants put up nearly 70,000 pounds of their own money to finance the venture, and the East India Company was born.

The corporation relied on a "factory" system, leaving representatives it called "factors" behind to set up trading posts and allowing them to source and negotiate for goods. Thanks to a treaty in 1613 with the Mughal emperor Jahangir, it established its first factory in Surat in what is now western India. Over the years, the company shifted its attention from pepper and other spices to calico and silk fabric and eventually tea, and expanded into the Persian Gulf, China, and elsewhere in Asia.

The East India Company's royal charter gave it the ability to "wage war," and initially it used military force to protect itself and fight rival traders. In 1757, however, it seized control of the entire Mughal state of Bengal. Robert Clive, who led the company's 3,000-person army, became Bengal's governor and began collecting taxes and customs, which were then used to purchase Indian goods and export them to England. The company then built on its victory and drove the French and Dutch out of the Indian subcontinent.

In the years that followed, the East India Company forcibly annexed other regions of the subcontinent and forged alliances with rulers of territory they could not conquer. At its height, it had an army of 260,000 (twice the size of Britain's standing army) and was responsible for almost half of Britain's trade. The subcontinent was now under the rule of the East India Company's shareholders, who elected "merchant-statesmen" each year to dictate policy within its territory.

But financial woes and a widespread awareness of the company's abuses of power eventually led Britain to seek direct control of the East India Company. In 1858, after a long wind down, the British government finally ended company rule in India. By 1874, the company was a shell of its former shelf and was dissolved.

By then, the East India Company had been involved in everything from getting China hooked on opium (the Company grew opium in India, then illegally exported it to China in exchange for coveted Chinese goods) to the international slave trade (it conducted slaving expeditions, transported slaves, and used slave labor throughout the 17th and 18th centuries). The East India Company may have since been overshadowed by modern capitalism, but its legacy is still felt around the world.

Influence of the Apprentice System

Aside from the East India Company, businesses, in most cases, were run by a single proprietor or a couple of partners. Management was tyrannical in nature—largely because of the ingrained apprentice system, which rendered employees helpless to disobey or quit.

According to *Encyclopedia Britannica*:

From the earliest times, in Egypt and Babylon, training in craft skills was organized to maintain an adequate number of craftsmen. The Code of Hammurabi of Babylon, which dates from the 18th century BCE, required artisans to teach their crafts to the next generation. In Rome and other ancient societies, many craftsmen were slaves, but, in the later years of the Roman Empire, craftsmen began to organize into independent collegia intended to uphold the standards of their trades.

By the 13th century, a similar practice had emerged in western Europe in the form of craft guilds. Guild members supervised the product quality, methods of production, and work conditions for each occupational group in a town. The guilds were controlled by the master craftsmen, and the recruit entered the guild after completing his training as an apprentice—a period that commonly lasted seven years. It was a system suited to domestic industry, with the master working in his own premises alongside his assistants. This created something of an artificial family relationship, in that the articles of apprenticeship took the place of kinship.

As time went on, however, governments had to contend with the exclusionary practices of the guilds, whose members could monopolize their trades in each town. Powerful guilds, for example, could levy high fees against outsiders to prevent them from entering a trade. Even apprenticeships could be restricted, with preference given to the sons of guild members or the sons of wealthy acquaintances. Responding to these improprieties, the English government tried to define the conditions of apprenticeship with the Statute of Artificers of 1563, which attempted to limit exclusionary practices and to ensure adequate labor.

The notion of individual training extended beyond the craft guilds in the Middle Ages. For example, universities advanced the same principle with the master's degree, as did religious orders that required newcomers to pass through a novitiate. In medicine, the guild system applied to the surgeon, who also acted as barber and was regarded as a craftsman with less prestige than the physician. Lawyers served an apprenticeship by working in close association with a master of the profession.

The Industrial Revolution altered attitudes toward training. Machines created a need for both skilled workers (such as machinists or engineers) and unskilled workers. Unskilled workers who showed aptitude advanced to semiskilled jobs. Apprenticeships actually grew in importance with the development of trade unions, which were created to uphold quality and control recruitment (by protecting union jobs).

In England, apprenticeship was maintained by the craft industries and even extended to analogous fields. The education system, for example, offered various apprentice programs for student teachers, and there was a comparable system of training for young farmers.

Apprenticeship was fairly common in the American colonies, with indentured apprentices arriving from England in the 17th century. (Benjamin Franklin served as apprentice to his brother in the printing trade.) But, apprenticeship in colonial America was less important than in Europe because of the high proportion of skilled workers in the colonies.

Because modernization and industrialization brought new impetus to the division of labor, the development of large-scale machine production increased the demand for workers with specialized skills. The more ambitious among them sought to increase their effectiveness and potential for advancement by voluntary study. To meet this need, mechanics' institutes were established, such as the one founded in London in 1823 by George Birkbeck, which still exists as Birkbeck College, and Cooper Union for the Advancement of Science and Art in New York City, established in 1859. In France, technical education on a national scale dates from 1880.

The American Enterprise

American corporations first appeared in the late 18th century and almost immediately became key drivers of the inchoate economy. Although the new company format also had existed in England and Holland, the US quickly began to lead the world in corporate development.

Small banking corporations had existed in the first years after the Revolutionary War. The first notable industrial enterprise, however, was the Boston Manufacturing Company, which imported its textile business model from Great Britain.

The corporate structure and easy access to capital, in the 1820s, catalyzed the American Industrial Revolution.

During the "Gilded Age," as Mark Twain labeled the second half of the 19th century, America became the world's greatest innovator and leading economic power—and saw the growth of railroads, the consumption of oil and electricity, and corporations, which were simple to form. Most states allowed free incorporation and required only a simple registration.

At the beginning of the 20th century, when Teddy Roosevelt was president, antitrust legislation put a temporary dent in corporate expansion, but it quickly rebounded.

Corporation structure began to evolve because of a new governance paradigm, increased government regulations, shareholder demands, foreign competition, and academic influence in finance and management.

After the Stock Market Crash of 1929 and the Great Depression, many Americans took a dim view of corporations.

Adolf Berle and Gardiner Means, in 1932, wrote *The Modern Corporation and Private Property,* opining that owners of corporations—shareholders—had too often ceded control to boards and management, allowing them to enrich themselves by operating their companies illegally. This book became the foundation for increased transparency and accountability.

In 1934, President Franklin D. Roosevelt established the Securities and Exchange Commission (SEC): to protect investors; maintain fair, orderly, and efficient marketplaces; and facilitate capital formation. The SEC requires public companies and other regulated companies to submit quarterly and annual reports, and other periodic reports. Also, company executives must provide a narrative account, called the "management discussion and analysis" (MD&A), that outlines the previous year of operations and explains how the company fared in that time period. The MD&A also touches on the upcoming year, outlining future goals and approaches to new projects. In an attempt to level the playing field for all investors, the SEC maintains an online database called EDGAR (the Electronic Data Gathering, Analysis, and Retrieval system) from which investors can access this and other information filed with the agency.

At the end of World War II (1945), America entered a period of meteoric growth. American corporations dominated the world, and the public favored them.

In the 1960s, the Penn Central Railway diversified into pipelines, hotels, industrial parks, and commercial real estate. Conglomerates didn't work. The company filed for bankruptcy in 1970. In 1974, the SEC sued three outside directors for misrepresenting the company's financial condition and Penn Central executives for a wide range of misconduct.

In 1976, the term "corporate governance" appeared for the first time in the Federal Register, the official journal of the federal government, and corporations started to form audit committees and appoint more outside directors. The NYSE began requiring each listed corporation to have an audit committee composed of all independent board directors.

In the 1980s, Japanese and German corporations emerged as bonafide competitors to American companies.

And, the US, under President Reagan, became politically conservative, precipitating an end to corporate-governance reform and opposition to deregulation.

In the 1980s, the "Deal Decade," institutional shareholders grabbed more shares, boosting their control. They stopped selling out when times got tough. Executives went on the defensive and struck deals to prevent hostile takeovers. Carl Icahn became known as a "corporate raider" because of his hostile takeovers like that of TWA, which he controlled in 1985 and subsequently dismantled.

In 2002, Congress passed the Sarbanes-Oxley Act, (SOX) which President George W. Bush signed into law. SOX required top execs, individually, to certify the accuracy of financial information, and it increased the oversight role of boards of directors and the independence of the outside auditors who review the accuracy of corporate financial statements. SOX was a reaction to corporate and accounting scandals, including those affecting Enron, Tyco International, Adelphia, Peregrine Systems, and WorldCom. These scandals cost investors billions of dollars when the share prices of the affected companies plummeted.

By 2007, investment banks were taking excessive risks, igniting a possible collapse of global financial systems. To wit: the 2008 demise of Lehman Brothers begat the worst international banking crisis since the Great Depression.

So, in 2010, the US Congress passed the Dodd-Frank Wall Street Reform and Consumer Act to promote financial stability and contain and limit the effect of the next financial crisis. President Obama signed it into law.

Wokeness Is the New Governance

Changes in corporate governance and public opinion of corporations vacillate with the times—based on the political party in power, scandals du jour, and university propaganda.

Today, universities teach students to be perpetually offended, aggrieved, and outraged; to protest; to hate America, Israel, capitalism, individual liberty, men, straights, whites, competition, conservatives, meritocracy, and Christians. All of it well-documented, influencing how corporations are run—because their recruitment pool comes from these universities.

Following the murder by a Minnesota cop of George Floyd, on May 25, 2020, there were widespread protests and Black Lives Matter (BLM)/Antifa-instigated riots, looting, and murders across America. In fearful, *emotional* response, to show "solidarity" with the mob, mollify their employees, and prevent or minimize looting of *their* buildings, woke CEOs of Airbnb, Apple, Amazon, Bank of America, Ben & Jerry's, Dropbox, Cisco, CitiGroup, Comcast, Disney, IBM, Intel, Netflix, NFL, Nike, Pepsi, PNC, Uber, UBISOFT, and many other firms rushed to support the violent BLM with millions of dollars in cash donations; Twitter hashtags; corporate-website banners; plaintive emails about eliminating **nonexistent systemic racism** to shareholders, employees, and customers.

That corporate cash belongs to the shareholders, who may not know that the BLM Foundation *advocates defunding police and abolishing prisons.* What will happen when armed thugs and released prisoners replace the defunded police? They'll come after those CEOs, where they live. Duh.

The ever-pandering Joe Biden blatantly lied during his presidential campaign, claiming that America, *the country that twice elected Obama and made Oprah Winfrey a billionaire,* is based on "systemic racism."

In fact, according to the Bureau of Justice Statistics, US Department of Justice, *over 90 percent of blacks are killed by other blacks*—not by white cops. Are there rogue cops? Yes, a tiny percentage of the total. But, there is no *systemic* racism in America.

Famed black economist Dr. Thomas Sowell, senior fellow on public policy at the Hoover Institution of Stanford University in California, opined to Mark Levin on Fox News, on July 12, 2020, about "systemic racism" in America:

> "It [systemic racism] really has no meaning that can be specified and tested in the way that one tests hypotheses. It does remind me of the propaganda tactics of Joseph Goebbels during the age of the Nazis, in which he is supposed to have said that people will believe any lie if it's repeated long enough and loud enough, and that's what we're getting."

Nevertheless, American CEOs either believe or pretend to believe Biden's lie, which the mainstream media happily perpetuate and reinforce. Facts don't matter.

On June 3, 2020, Senator Tom Cotton (R-AL) penned an op-ed piece in the *New York Times* about the need to use military force to quell rioting across America. The NYT had invited him to do so, to show some political balance. But, many childish NYT readers and writers objected to *any* point of view other than their progressive one. So, publisher A.G. Sulzberger fired the scapegoated editor of the editorial page, James Bennet. Once again, the inmates run the asylum.

On July 14, 2020, Bari Weiss, facing hatred from the internal leftist mob, resigned her post as op-ed staff editor and culture-politics writer at the *New York Times*. Here are selected, unedited quotes from her resignation letter:

> "Instead, a new consensus has emerged in the press, but perhaps especially at this paper: that truth isn't a process of collective discovery, but an orthodoxy already known to an enlightened few whose job is to inform everyone else."

> "Twitter is not on the masthead of the *New York Times*. But, Twitter has become its ultimate editor. As the ethics and mores of that platform have become those of the paper, the paper itself has increasingly become a kind of performance space. Stories are chosen and told in a way to satisfy the narrowest of audiences, rather than to allow a curious public to read about the world

and then draw their own conclusions. I was always taught that journalists were charged with writing the first rough draft of history. Now, history itself is one more ephemeral thing molded to fit the needs of a predetermined narrative."

"My own forays into Wrongthink have made me the subject of constant bullying by colleagues who disagree with my views. They have called me a Nazi and a racist; I have learned to brush off comments about how I'm 'writing about the Jews again.' Several colleagues perceived to be friendly with me were badgered by coworkers. My work and my character are openly demeaned on companywide Slack channels, where masthead editors regularly weigh in. There, some coworkers insist I need to be rooted out if this company is to be a truly 'inclusive' one, while others post ax emojis next to my name. Still, other *New York Times* employees publicly smear me as a liar and a bigot on Twitter, with no fear that harassing me will be met with appropriate action. They never are."

Welcome to present-day corporate governance, where virtue-signaling supplants leadership and the SEC, and woke kowtowing trumps management by objective.

Alas, in such a world, intrabranding, and, therefore, branding, are and will be next to impossible to execute.

Is *your* company like the *New York Times*?

CHAPTER FIVE

Evolution of Employees & HR

Who's in Charge?

Full disclosure: Throughout my career, I never had any use for the HR (human resources) department. I always saw it as the "no" department. I made this clear in Chapter 7 of *Brand Is Destiny*:

> Many moons ago, right out of engineering school, I landed as a manufacturing engineer at the Eastman Kodak Company in Rochester, NY. After toiling awhile on a movie projector and an electronic flash for the short-lived instant camera, I began to contemplate my future. So, I ambled down to my least-favorite department, HR, where I expressed to the available puke my desire to pursue an MBA. I also requested confirmation that Kodak would pay for it.
>
> The HR puke pointed to a portrait on the wall and asked me to identify the man in the frame. "Walter Fallon," I replied. "And do you know his title?" he continued. "The CEO," I rejoined. "That's right. Walter

Fallon is the CEO, and he's a chemist. He didn't need an MBA to succeed here, and neither do you. Now, you can go get an MBA, and we'll probably pay for it, but I counsel against it."

His attitude stunned me. I walked back to my desk, dejected, wondering what kind of company I had joined.

By blowing me off, the snarky HR rep had revealed Kodak's "personality." Every company has one, and it usually stems from the CEO. In an old company like Kodak, the personality is ingrained via a series of same-thinking CEOs.

I shelved my plans to get an MBA for a few years until after moving to Boston and working for another company.

Kodak filed for Chapter 11 bankruptcy in 2012. The prominent [corporate] personality traits—*arrogance and insularity*—that effected this implosion were obvious when I worked there.

Things have changed a tad in the corporate world since I began my career: HR has gained *and* lost power, as have employees. It's a situation in constant flux.

To wit: Here's an excerpt of an article appearing in *Adweek*, on June 2, 2020, about an employee revolt at Facebook against CEO Mark Zuckerberg over one of President Trump's recent tweets:

Tensions are escalating inside Facebook over the social platform's laissez-faire approach to the president's posts.

While Twitter took an active approach to Donald Trump's account last week—including flagging a tweet that encouraged shooting unarmed protesters— Facebook chose to interpret Trump's message differently and has not modified the same post.

Some Facebook employees are upset over the policy— and tweeted about it.

"I work at Facebook and I am not proud of how we're showing up," Jason Toff, director of product management, tweeted early Monday morning. "The majority of coworkers I've spoken to feel the same way. We are making our voice heard."

Design manager Jason Stirman tweeted that he "completely disagrees" with Facebook CEO Mark Zuckerberg's decision to "do nothing about Trump's recent posts, which clearly incite violence."

"I'm not alone inside of FB," he added. "There isn't a neutral position on racism."

The *New York Times* reported today [06/02/20] that dozens of Facebook employees are also staging a virtual "walkout" to call out the social network's inaction over Trump's post encouraging violence against protesters.

"We recognize the pain many of our people are feeling right now, especially our Black community," a Facebook spokesperson told Adweek. "We encourage employees to speak openly when they disagree with leadership. As we face additional difficult decisions around content ahead, we'll continue seeking their honest feedback."

Zuckerberg authored a lengthy Facebook post Friday, saying he had a "visceral negative reaction to this kind of divisive and inflammatory rhetoric," but said he is responsible for reacting "as the leader of an institution committed to free expression."

Last Tuesday, after years of pressure, Twitter took unprecedented action against Trump's account, placing a fact-check label on two of his tweets about mail-in ballots. Trump responded by lashing out, accusing Twitter of interfering with the election and promising retribution.

He took it a step further Thursday when he signed an executive order that, while legally fraught, threatens social media companies like Twitter and Facebook by attempting to curb liability protections afforded by Section 230 of the Communications Decency Act.

By the end of the week, Trump had not cooled his rhetoric and Twitter didn't back down. With protests raging in Minneapolis and elsewhere in the country over the police killing of George Floyd early Friday morning, Trump sent a tweet with the quote "when the looting starts, the shooting starts."

Twitter promptly blurred out the tweet with a "public interest notice," defending the move by claiming it breaks site rules by "glorying violence." Still, the platform did not remove the tweet, and users can click to see it—because, it claimed, Trump's tweets are newsworthy as president.

Meanwhile, Facebook allowed the same post to stand unaltered on its site.

If anyone had tried revolting against the management at any company where I have worked, he would have been fired on the spot.

Seemingly, the employees are now in charge. Wrong. Speak or write anything that contradicts HR's policies on gender: you're fired. Here's an excerpt from *The Daily Wire* on January 21, 2020:

Only two sexes? Pshaw.

There's transgender, non-binary, genderfluid, agender, bigender, polygender, intersex, neutrois, androgyne, intergender, demigender, greygender, maverique, novigender, dyadic, and more. If all that confuses you, there's even something called "gender apathetic," defined by one website as "when you really do not identify nor care about any particular gender."

But a *Denver Post* columnist recently had the audacity to say there were just two sexes, male and female. And he says that got him fired.

Jon Caldara, president of the libertarian Independence Institute, announced on Facebook that he was canned by the *Post* after his column appeared.

In his piece, headlined "Colorado Dems should let sun shine on their hospital fees and sex-ed curriculum," Caldera makes a case for government transparency, especially in healthcare. Then the columnist weighs in on school curricula:

Democrats don't want transparency in hospital billing, and they certainly don't want education transparency when it comes to their mandate to convince your kid that there are more than two sexes, even if it's against your wishes.

Among the most controversial laws that passed last year was the comprehensive human-sexuality education mandate which ripped local control away from your neighborhood schoolboard. Now, if your school district wants to teach even basic sex-ed, the teacher must also teach the "health needs" of LGBT individuals.

And, in the anti-free-speech style that the left now embraces, the new law bans discussions that "employ gender stereotypes," or any language the state's new oversight board of LGBT activists deem *"stigmatizing."* In case you hadn't noticed, just about everything is stigmatizing to the easily triggered, perpetually offended.

He also notes that "some parents weren't thrilled a couple of years back when during school their little ones in Boulder Valley School District were treated to videos staring a transgender teddy bear teaching the kids how to misuse pronouns or when Colorado's 'Trans Community Choir' sang to kids about a transgender raven."

After saying in his Facebook post that he was fired, Caldera defends his piece.

"My column is not a soft-voiced, sticky-sweet, NPR-styled piece which employs the language now mandated by the victim-centric, identity politics driven media," he said in his post. "What seemed to be the last straw for my column was my insistence that there are only two sexes, and my frustration that to be inclusive of the transgendered (even that word isn't allowed), we must lose our right to free speech."

Caldera says he employs "plain talk that doesn't conform to the newspeak law of 'use only the words mandated by the perpetually offended.' So, it is labeled as 'mean-spirited' and banned. If conservatives and libertarians are granted a voice in the mainstream media, they must use the language of their ideological opponents. That is, they are not allowed to have their own voice."

After reading the two excerpts above, it is difficult to tell whether management or employees are in charge.

Evolution of HR

Pre-1900s to 1920

The link between worker wellbeing and productivity emerged between 1890 and 1920. The buzzwords then were industrial betterment, industrial welfare, and scientific management. The goals of these initiatives were employee stability and loyalty.

1920s–1950s

Era of accepting workers as humans, not mere robots. Personnel departments and manpower development increased in popularity. Companies instituted training programs and worked with labor unions to improve compensation (National Labor Relations Act became law in 1935).

1960s–1980s

The Myers-Briggs Type Indicator surfaced in 1962. The US Congress passed the Equal Pay Act of 1963 and Civil Rights Act of 1964. Employee motivation became the new HR priority. Organizational management and industrial psychology reinforced employees' needs for achievement, advancement, and recognition.

1990s–2020

IT and smartphone technologies helped morph HR from managing "personnel" to increasing employee engagement and retention, and buttressing corporate culture. Employees can work anytime, from anywhere. On the flip side, political correctness (PC) took over boardrooms and HR departments, dictating what employees could think, say, and do. "Diversity and inclusion" became a new priority, and many large corporations now have a chief diversity officer (CDO).

2020 and Beyond

SHRM (Society for Human Resources Management) asserts, "Tomorrow's HR leaders will need to be bigger, broader thinkers, and they'll have to be tech-savvy and nimble enough to deal with an increasingly *agile* and restless workforce."[Emphasis added.]

Here's the reality in numerous enterprises: Because of "woke" political correctness, virtue-signaling, and restrictions on thought, speech, and actions, employees are often so petrified to speak up, to opine honestly, to criticize the shoddy work of protected peers and subordinates—women, gays, transgenders, blacks, Hispanics, etc.—that they keep quiet. This silence hurts them, their employers, their customers, and, ultimately, their employers' shareholders.

CHAPTER SIX

The Power of Agility

Think Fast, Act Faster

Those who are agile can perceive, process, and proceed quickly, in real time. The roots of agility are *urgency, alignment, and communication.*

Imagine a pitcrew, responsible for keeping a racecar in the race, moving like a snail. You can't. Speed and accuracy are everything. When the car leaves the track and enters the pit for refueling and repairs, the team members, who have been watching the car circle the track, must quickly get the driver's feedback, communicate with each other, diagnose the problems, and fix those problems—*at lightning speed.*

If only companies would execute like pitcrews.

Usually, they don't. Except when virtue-signaling and wokeness—such as bowing down to social-justice mobs—are on the menu (suddenly killing Uncle Ben's, Aunt Jemima, Mrs. Butterworth, Eskimo Pie), then companies move with astounding swiftness. Otherwise, they inch along like snails.

Snails can't brand, can't win. They're stuck in time.

This, Gentlemen, Is War

Ford vs. Ferrari, starring Matt Damon as Carroll Shelby, racecar-designer extraordinaire, and Christian Bale as Ken Miles, champion racecar driver, is based on a true story. It's a thrilling movie and a supreme illustration of how agility, or the lack thereof, can make or break an enterprise

In 1965, Lee Iacocca, a VP at Ford, tried to alarm his peers and "The Deuce" (CEO Henry Ford II) at a C-suite meeting, where they were lamenting a huge sales slump. He opined that Ford had been wrong in its thinking and needed an injection of some Ferrari-style branding. Incredulous, the attendees wanted to know why big Ford should emulate a little Italian company that builds few cars:

> Iacocca: "*Why?* Because he [Enzo Ferrari] built the most cars? No. Because of what his cars *meant.* Victory. When Ferrari wins at Le Mans, people want

some of that victory. What if the Ford badge meant victory? And meant it where it counts: among the first group of 17-year-olds in history with money in their pockets." [Emphasis added.]

Iacocca then proposed that Ford merge with Ferrari and led a delegation of Ford execs to Italy, where Enzo Ferrari rejected Ford's offer (eventually, Fiat bought Ferrari).

Via Iacocca, Enzo Ferrari relayed a message to The Deuce: Tell him that "Ford makes ugly little cars in an ugly factory. And, its executives are sons of whores." And, the Deuce is "fat, pig-headed, and not Henry Ford—just Henry Ford... the second."

Back at Ford HQ, when The Deuce hears these insults from Iacocca, he reacts *emotionally*:

"I want the best engineers, the best drivers. I don't care what it costs. We're gonna build a racecar, and we're gonna bury that greasy, good-for-nothing, devious wop a hundred feet deep under the finish line at Le Mans. And, I will be there to watch it. This isn't business. This, gentleman, is war."

Ford approached Carroll Shelby, a successful designer and builder of racecars, to help achieve Ford's mission to win at Le Mans by building a new car—in 90 days. Shelby accepts and enlists his friend—champion driver, mechanic, straight-talking Ken Miles, who, over a cup of coffee, schools Shelby in a corporate agility lesson:

"I can see you now sitting in the boardroom in Detroit, in your stripey overalls, with [Shelby's employees] Pops, Burner, and Chuck—a bunch of hotrodders, beatniks, speedfreaks. Have you been to Detroit? They have whole floors of lawyers. A million marketing guys. And they'll all line up to kiss your arse, get their photo taken with the great Carroll Shelby, and they'll head on back to their nice offices and dream up new ways to screw you. Why? Because they can't help it. They all just want to please their boss, who just wants to please his boss, who just wants to please HIS boss. And, they hate themselves for it, but deep down who they really hate is guys like you. Because you're not like them. You don't think like them. You're different."

After lots of hiccups, with mechanical and personnel problems and embarrassments on the track, and tiring of Ford's many layers of bureaucracy and red tape, Shelby gets a meeting with The Deuce in his office. While sitting in Ford's waiting room, Shelby had noticed a red folder being passed around and then handed to Ford's secretary, who places it on Ford's desk. Finally, Shelby enters Ford's office:

Ford II: "Shelby. Give me one reason why I don't fire everyone associated with this abomination, starting with you."

Shelby: "Well, sir, I've been thinking of that very question as I sat out in your lovely waiting room. And, while I was sitting there, I watched that little red folder

right there go through five pairs of hands before it got
to your mitts. And, that's not including the 22 Ford
employees who must've poked at it before it got to the
19th floor. With all due respect, sir, you can't win a
race by committee. It's like trying to run with a load in
your pants. You need one man in charge."

Iacocca looks at Shelby in total incredulity (nobody
talks to The Deuce like this!)

Shelby continues: "The good news is, if you ask me,
even with the extra weight in the trunk, we still
managed to put ol' Ferrari right where we want him."

Ford: "Did we?"

Shelby: "Oh, yes."

Ford: "Expand."

Shelby: "Well, sure, we haven't worked out how to
corner. Or, stay cool. Or, stay on the ground. And, a
lot of stuff broke. In fact, the only thing that didn't
break is the brakes. Right now, we don't even know if
the paint lasts 24 hours. But, our last lap clocked 218
MPH down the Mulsanne straight. Now, in all his years
of racing, Enzo ain't seen nothing move that fast. And,
now, he knows we're faster than him. Even with the
wrong driver [Ford execs had rejected Ken Miles as
their driver, because he wasn't compliant, but Shelby
had him reinstated—and he won, big time.] and the
committees. That's what he's [Enzo Ferrari] sitting in

Modena thinking about, right now. He's worried this year you might actually be smart enough to give me the control I need to win. So, yes, I'd say you got Ferrari exactly where you want him. You're welcome."

Iacocca's jaw dropped. Beebe [senior EVP] stared, horrified, at Shelby's insolence. Ford eyeballed Shelby for a good seven seconds. He stood. Looked out the window. A crease of a smile.

Ford: "See that little building down there? In WW2, three out of five US bombers rolled off that line. You think Roosevelt beat Hitler? Think again. This isn't the first time Ford Motor's gone to war in Europe. *We know how to do more than push paper.* And—(points to himself)—there is *one man* running this company. And, you report to *him*. Go ahead, Shelby. Go to war." [Emphasis added.]

Despite lots of wrangling (Beebe stepped in, in Ford's stead, to supervise Shelby, who mostly ignored him), Shelby and Miles won Le Mans in 1966. Ford Motor also won it in 1967, 1968, and 1969.

Shelby taught The Deuce that agility is *everything*: "You can't win a race by committee." Without Shelby's brains, wile, chutzpah, laser-like focus, and leadership, Ford Motor couldn't have and wouldn't have succeeded at Le Mans, especially in that compressed timeframe.

Does your enterprise manage by "red folders"?

Finally, how did that success at Le Mans rub off on the rest of the company? According to the chart below, there's no evidence that it did. Ford still operated through layers of stifling red-folder bureaucracy, made the same mediocre cars, and didn't appreciably move the needle vs. General Motors.

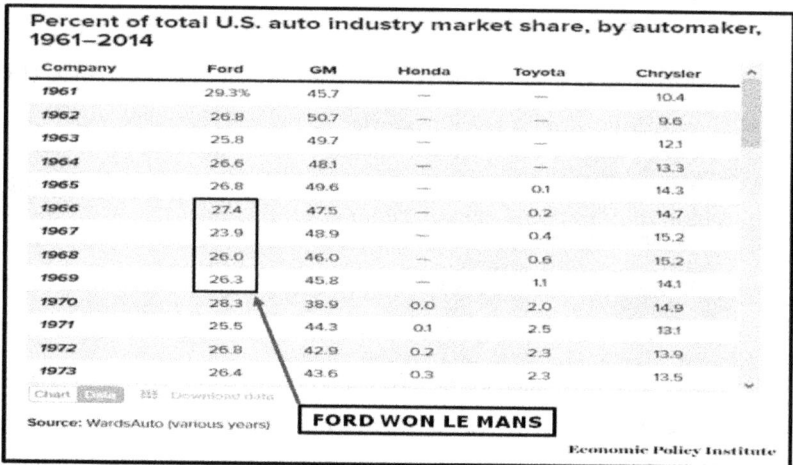

Percent of total U.S. auto industry market share, by automaker, 1961–2014

Company	Ford	GM	Honda	Toyota	Chrysler
1961	29.3%	45.7	—	—	10.4
1962	26.8	50.7	—	—	9.6
1963	25.8	49.7	—	—	12.1
1964	26.6	48.1	—	—	13.3
1965	26.8	49.6	—	0.1	14.3
1966	27.4	47.5	—	0.2	14.7
1967	23.9	48.9	—	0.4	15.2
1968	26.0	46.0	—	0.6	15.2
1969	26.3	45.8	—	1.1	14.1
1970	28.3	38.9	0.0	2.0	14.9
1971	25.5	44.3	0.1	2.5	13.1
1972	26.8	42.9	0.2	2.3	13.9
1973	26.4	43.6	0.3	2.3	13.5

Chart **Data** ::: Download data

Source: WardsAuto (various years)   **FORD WON LE MANS**

Economic Policy Institute

Ford *did* prove that, where there's a will, there's a way. But, the will *must* be to better serve customers, not use the shareholders' capital to execute a vendetta against one man.

Nevertheless, that proof should serve as an important lesson about agility to all CEOs, in every industry.

Intrabranding is the keystone of corporate agility because when execs and employees row in the same direction, in the right direction (purpose), and speak the same language, the enterprise can succeed in spades.

CHAPTER SEVEN

Entroprise Basics

Effective intrabranding requires an aligned enterprise. Too many, sadly, are dysfunctionally *unaligned*: productive communication is nonexistent.

I call this organization an *entroprise*.

A portmanteau of entropy (chaos) and enterprise, I coined and introduced *entroprise*, a chaotic enterprise, in Chapter 7 of *Brand Is Destiny*, my previous book.

Below are five key attributes of the typical entroprise:

THE ENTROPRISE

DEFICIENT/DYSFUNCTIONAL

- **Grasp of Customers' Lives**
- **Purpose & Direction (brand)**
- **Employee Urgency & Accountability**
- **Critical Thinking**
- **Communication & Collaboration**

© 2020 MarcRudov.com

Without a grasp of customers' business and personal lives, your company can't have a discernible purpose and direction (brand). If your employees don't know why your company exists and where it's headed, they'll feel no urgency about anything and will behave without accountability.

Because public schools brainwash kids with socialism (California schools now teach that capitalism is racist), and many companies sheepishly and desperately try to outwoke each other, critical thinking is dead—ruining competitiveness,

Finally, defective discourse, which begets the four other entroprise attributes, is the rule, not the exception in the corporate world.

Caveat: Even if a company can easily convey a missive, to be a *message*, it must *move* people—internally and externally. For example, Joe Biden, palpably senile, does daily webcasts, repeated by all media outlets. But, he can't form complete sentences, can't move anyone, and, thus, has no message. Conveying without *convincing* is merely palavering.

People Are Lousy Communicators

Why do the bulk of relationships and marriages fail? Poor communication. Generally speaking, people come from families in which crystal-clear, honest communication is rare. Ask them. How, then, do they go into their workplaces and magically become good communicators? They don't.

Unless management stops, admonishes, and retrains, them, they continue their dysfunctional at-home proclivities and behaviors in their workplaces: lying, hiding, cheating, snubbing, retaliating, humiliating, withdrawing, and passive-aggressive sabotaging.

Is it any wonder that companies have communication and collaboration problems?

Lack of Urgency & Accountability

A few years ago, I received a call from a distressed Fortune 50 executive, lamenting that his company was struggling to create a "culture of accountability."

Newsflash: If your employees exhibit "Who, me?" attitudes, your company lacks accountability. This behavioral deficit adds cost to every process, impairs product quality, angers customers, repels investors, and makes intrabranding impossible.

Accountability is a value taught (or mocked) in homes, neighborhoods, schools, houses of worship; in music, on TV, in movies; and from politicians, from business leaders, and in the general culture.

If children aren't raised and trained to be accountable—to do what's necessary and proper, to take responsibility for their actions, without being told—it's harder to expect and enforce such behavior in adulthood.

Moreover, children who witness the hypocrisy of unaccountable adults who misbehave with impunity will disregard everything they're taught.

In corporations, accountability starts at the top and propagates downward.

When employees know that their CEO, and all the other execs on the org chart, will enforce both carrots and sticks, they will toe the line. If, on the other hand, the CEO encourages and rewards cheating, that will be the result.

Lou Holtz, Hall of Fame football coach and popular motivational speaker, spoke in 2019 with sportscaster Joe Buck about his life, career, and success philosophy. Holtz compared athletes of today with those of 40 years ago:

"Today, everybody wants to talk about their rights and their privileges. Forty years ago, we talked about our *obligations and our responsibilities.* I believe that we still need to get back to the obligations and responsibilities you have to other people."

Lou Holtz brilliantly defined accountability. By shining a light on spoiled athletes, he explained its absence in today's culture. The attitude of entitlement, taught in schools and blanketing our society, is the antithesis of accountability.

Yet, Holtz downplayed the situation: He didn't mention political correctness, safespaces, microaggressions, perpetual outrage, and the other hallmarks of our Millennial-driven culture that obviate accountability.

CEOs must create accountable companies from a large pool of unaccountable, entitled employees, influenced and nourished by a culture that neither demands nor reinforces obligation and responsibility.

Accountability requires leadership, starting with *you.*

If you want to run an accountable company with a strong brand, you need people who live and breathe obligation and responsibility—and demand it from those around them. This is an uphill fight.

Don't allow "Who, me?" slackers into your company. Retrain those who slipped in through faulty interviewing practices, and purge the rest.

Fire Your Easily Offended

Dr. Jeremy B. Bernerth, business professor and social scientist at San Diego State University, published, in May 2020, a study in the *Journal of Business Research* about the effects in the workplace of people who are easily offended— those with a high *proclivity to be offended* (PTBO). He drew an unsurprising conclusion: *the easily offended make terrible employees, because they never get anything done.*

Bernerth asked almost 400 employees, aged 25.9 on average, across seven US colleges, about different events that received substantial media attention: nine PTBO items, eight moral-outrage items, 11 microaggression items, and nine political-correctness items.

Bernerth claims that some people have a high PTBO— a state-like tendency to be sensitive to and offended by customarily innocuous societal events and traditions, like the playing of America's National Anthem and the term "Islamic terrorist," deeming them in violation of moral or equitable standards. Basically, they're easily triggered by anything.

The study also indicates that the easily offended are less concerned with helping others, despite constant virtual-signaling that altruism is their primary goal.

Finally, Bernerth links the high PTBOs (perpetually outraged Millennials and Gen-Zers) to poor task performance, bad citizenship, counterproductive work behavior, high job dissatisfaction, and low engagement. Notwithstanding this, woke orgs don't grasp how to vet applicants and employees who express indignation at an array of events and traditions, and, consequently, hire them—*a deathknell to intrabranding*

Lack of Critical Thinking

Every year, freshmen and seniors in the USA take the Collegiate Learning Assessment Plus (CLA+), issued by the non-profit Council for Aid to Education. Its goal is to measure critical-thinking skills of students entering and leaving 200 institutions of "higher learning."

The results in 2017 are disastrous. The College Fix summarized the *Wall Street Journal's* analysis:

> "The Journal found that at about half of schools, large groups of seniors scored at basic or below-basic levels," according to *Newsweek.* For many of these seniors, this means that "they can generally read documents and communicate to readers but can't make a cohesive argument or interpret evidence."

Let's repeat this astounding conclusion: *Many seniors cannot make a cohesive argument or interpret evidence.*

Parrot, React, and Follow

Should anyone be surprised? As you can read in *Brand Is Destiny*, universities and colleges are teaching students not to think, not to question, not to analyze—just to parrot, react, and follow.

Conclusion: Our so-called education system, from K through college, is an expensive, unmitigated catastrophe.

On June 19, 2020, uneducated, anarchist morons in San Francisco toppled the statue of Ulysses S. Grant, our 18th president, who won the civil war and freed the slaves. Grant, a Republican, also snuffed out the Democrat-originated KKK and sent soldiers to the South to enforce Reconstruction. Will these idiots be your employees someday?

Critical thinking, the objective analysis of facts to form a judgment, is discouraged at every level—from gender fluidity to climate science to food—by the government, the media, the universities, and the mobocracy.

The fascist mob in Portland forced two white women to close their burrito truck—because they were appropriating Mexican culture. In a free society, nobody even fought back. Should this mean that Mexicans aren't "allowed" to make hamburgers, either?

Remember when Barack Obama oxymoronically and wrongly told us that climate change is settled science?

The late Charles Krauthammer wrote in a 2014 article: "There is nothing more anti-scientific than the very idea that science is settled, static, impervious to challenge."

Albert Einstein, via his Theory of Relativity, historically *unsettled* the science of his day.

But, today's brainwashed, robotic students, burdened by safespaces, microaggressions, and nonstop protesting, know nothing about objective analysis and thinking.

Unhinged snowflakes rioted at Evergreen College, in Olympia, Washington, because a white professor refused to leave campus on *anti-white day*. The feckless administration capitulated. **Lesson learned:** thinking, debate, freedom, maturity, and the Constitution have no place in our society.

These unthinking, unchallenged, coddled students are now joining the corporate ranks. Because political correctness is rampant in the modern enterprise, which practically worships them, critical thinking has become scarcer, and competitiveness is suffering. Adults have suspended logic to accommodate spoiled children.

Note: A decision devoid of critical thinking is defective and will backfire.

Example: The board's lack of critical thinking allowed former CEO Marissa Mayer to run Yahoo into the ground. In 2008, Microsoft tried to buy Yahoo for $45 billion. Instead, the board dawdled for four years, then hired Ms. Mayer in 2012. For the next five years, she made one blunder after another—and even tried to purge male employees—with impunity.

Why did Yahoo's board fail to stop Mayer? Political correctness? Fear of sexism accusations, perhaps? In June of 2017, Verizon purchased Yahoo's core Internet assets for $4.5 billion; Mayer departed with at least $186 million.

Political correctness and critical thinking are mutually exclusive. Sadly, political correctness now trumps critical thinking: in the media, in our schools, in our federal and local governments, and in the corporate world.

Employees who, because of political correctness (PC) or inabilities to make cohesive arguments and interpret evidence, will communicate dysfunctionally (if at all) and turn your enterprise into an entroprise, akin to a hole-laden hunk of Swiss cheese, thereby crippling its effectiveness.

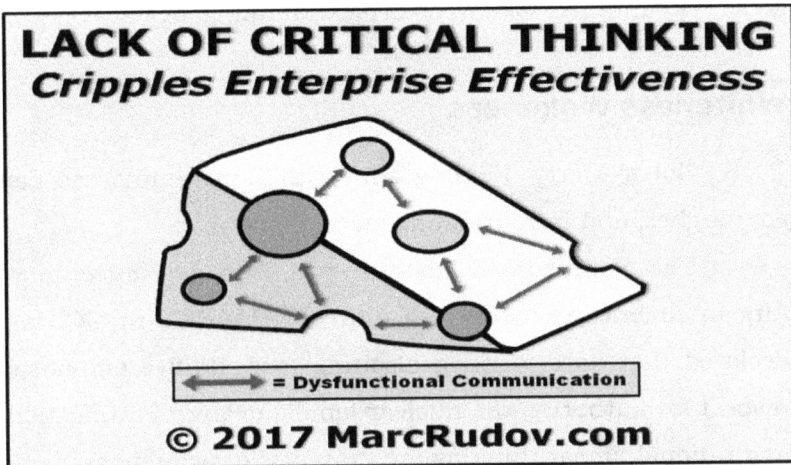

LACK OF CRITICAL THINKING
Cripples Enterprise Effectiveness

= Dysfunctional Communication

© 2017 MarcRudov.com

That's right: Every poor communicator is a hole in your organization.

How can you expect to create a strong, unique brand, which sets your company's purpose and direction, if said company resembles Swiss cheese?

Generals and soldiers incapable of critical thinking will die and lose wars.

Enterprises incapable of critical thinking will lose customers and disappear. Think Sears.

If your employees fear criticizing each other's ideas and tasks—because of race, gender, ethnicity, religion, country of origin, or other reasons—expect dysfunctional communication and failure.

Never hire, promote, or seek the advice of those who can't analyze and challenge trends, make cohesive arguments, or interpret evidence—even if corporate PC pressures you to do so. The ultimate cost of no critical thinking: defeat.

Whiteness Wokeness

Not to worry. Finding critical-thinking employees just got tougher, and intrabranding more impossible.

The Smithsonian Institute-run National Museum of African American History & Culture (NMAAHC), in DC, has declared that hard work, goalsetting, quantitative emphasis, respect for authority, the nuclear family, delayed gratification, and rational, linear thinking are examples of "whiteness."

Think I'm joking? Look at the "White Culture" graphic on the next page, from the website of the Smithsonian, which in FY 2020 got $1 billion in taxpayer funding. The backlash was so severe that Smithsonian removed it within two days.

We'll find this Marxist propaganda in public schools. It will further lower the work ethics of your current and future employees. Such insulting trash, government-funded in this case, promotes failure. It's the *real* systemic racism.

TALKING ABOUT RACE | NMAAHC

ASPECTS & ASSUMPTIONS OF WHITENESS

WHITE CULTURE IN THE UNITED STATES

White dominant culture, or **whiteness**, refers to the ways white people and their traditions, attitudes and ways of life have been normalized over time and are now considered standard practices in the United States. And since white people still hold most of the institutional power in America, we have all internalized some aspects of white culture — including people of color.

Rugged Individualism
- The individual is the primary unit
- Self-reliance
- Independence & autonomy highly valued + rewarded
- Individuals assumed to be in control of their environment, "You get what you deserve"

Family Structure
- The nuclear family: father, mother, 2.3 children is the ideal social unit
- Husband is breadwinner and head of household
- Wife is homemaker and subordinate to the husband
- Children should have own rooms, be independent

Emphasis on Scientific Method
- Objective, rational linear thinking
- Cause and effect relationships
- Quantitative emphasis

History
- Based on Northern European immigrants' experience in the United States
- Heavy focus on the British Empire
- The primacy of Western (Greek, Roman) and Judeo-Christian tradition

Protestant Work Ethic
- Hard work is the key to success
- Work before play
- "If you didn't meet your goals, you didn't work hard enough"

Religion
- Christianity is the norm
- Anything other than Judeo — Christian tradition is foreign
- No tolerance for deviation from single god concept

Status, Power & Authority
- Wealth = worth
- Your job is who you are
- Respect authority
- Heavy value on ownership of goods, space, property

Future Orientation
- Plan for future
- Delayed gratification
- Progress is always best
- "Tomorrow will be better"

Time
- Follow rigid time schedules
- Time viewed as a commodity

Aesthetics
- Based on European culture
- Steak and potatoes; "bland is best"
- Woman's beauty based on blonde, thin — "Barbie"
- Man's attractiveness based on economic status, power, intellect

Holidays
- Based on Christian religions
- Based on white history & male leaders

Justice
- Based on English common law
- Protect property & entitlements
- Intent counts

Competition
- Be #1
- Win at all costs
- Winner/loser dichotomy
- Action Orientation
- Master and control nature
- Must always "do something" about a situation
- Aggressiveness and Extroversion
- Decision-Making
- Majority rules (when Whites have power)

Communication
- "The King's English" rules
- Written tradition
- Avoid conflict, intimacy
- Don't show emotion
- Don't discuss personal life
- Be polite

The Saboteur

Intrabranding is challenging enough, as per the previous sections of this chapter. Sabotage from disgruntled employees raises the stakes exponentially. It can occur inside the company or via badmouthing on social media and leaking confidential information to competitors or the press.

After sabotage at Tesla, Elon Musk sent this email, on June 17, 2018, to all his employees:

I was dismayed to learn this weekend about a Tesla employee who had conducted quite extensive and damaging sabotage to our operations. This included making direct code changes to the Tesla Manufacturing Operating System under false usernames and exporting large amounts of highly sensitive Tesla data to unknown third parties.

The full extent of his actions are not yet clear, but what he has admitted to so far is pretty bad. His stated motivation is that he wanted a promotion that he did not receive. In light of these actions, not promoting him was definitely the right move.

However, there may be considerably more to this situation than meets the eye, so the investigation will continue in depth this week. We need to figure out if he was acting alone or with others at Tesla and if he was working with any outside organizations.

As you know, there are a long list of organizations that want Tesla to die. These include Wall Street short-sellers, who have already lost billions of dollars and stand to lose a lot more. Then there are the oil & gas companies, the wealthiest industry in the world — they don't love the idea of Tesla advancing the progress of solar power & electric cars. Don't want to blow your mind, but rumor has it that those companies are sometimes not super nice. Then there are the multitude of big gas/diesel car company competitors. If they're willing to cheat so much about emissions, maybe they're willing to cheat in other ways?

Most of the time, when there is theft of goods, leaking of confidential information, dereliction of duty or outright sabotage, the reason really is something simple like wanting to get back at someone within the company or at the company as a whole. Occasionally, it is much more serious.

Please be extremely vigilant, particularly over the next few weeks as we ramp up the production rate to 5k/week. This is when outside forces have the strongest motivation to stop us.

If you know of, see or suspect anything suspicious, please send a note to [email address removed for privacy] with as much info as possible. This can be done in your name, which will be kept confidential, or completely anonymously.

Looking forward to having a great week with you as we charge up the super exciting ramp to 5000 Model 3 cars per week!

Will follow this up with emails every few days describing the progress and challenges of the Model 3 ramp.

Thanks for working so hard to make Tesla successful,

Elon

Sabotage is disruptive and costly. Don't hire angry, rebellious, protest-oriented, sullen, uncooperative employees.

Acorn of the Entroprise

It is illustrative to diagnose just how entroprises are formed. Below is an explanation adapted from Chapter 7 of *Brand Is Destiny*:

Rebecca Homkes, teaching fellow at London Business School, and MIT Sloan School senior lecturer Don Sull wanted to learn how well companies execute their strategies. The duo asked more than 11,000 senior executives in 400 firms to name their top-three priorities. Only about one-third could do so and, given five tries, only about half could agree on one.

These results don't surprise me. I was lecturing a group of CEOs and asked: *Were you to query your employees, at all levels, about why your company exists and how it is unique, what would they say?* The reaction from the room: blank stares, crickets, and palpable awkwardness.

First, the CEOs couldn't have answered my question: their companies were brandless. Second, intracompany communication is universally poor—and CEOs don't realize it.

CHALLENGE: Ask *your* employees the same question. Don't be surprised if they stumble and fumble for answers.

This unfamiliarity is akin to paratroopers landing at night in a foreign country—with no idea why, no knowledge of the terrain, and no clue about what each should do next. This is when "self-management" and chaos begin.

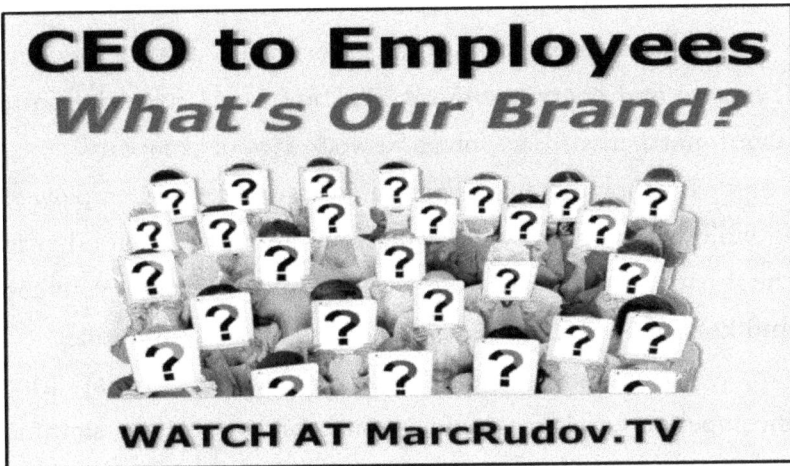

Recently, I bought a new iPhone at an Apple Store. The sales associate, while loading up my phone with the latest version of the iOS operating system, asked me what I do. I began answering him, explaining the gist of this book: *poor internal communication is the genesis of lousy external communication.* I asked him, What's Apple's brand? He gave me some generic boilerplate gobbledygook. I said, See, you don't know, and it's not your fault. It's Tim Cook's (Apple's CEO) fault. Apple's brand, I told him, is cool lifestyle.

I cannot reiterate my immutable axiom frequently enough: *With no brand, a murky brand, or a horrible brand— whether extant by design or default—your destiny shall be endless random drifting or hitting an iceberg. Either way, your bottom line is imperiled.*

Design or default? Yes. Some people work hard, with a touch of ignorance, arrogance, or defiance, to build a defective brand. Others achieve this ignoble outcome by doing nothing.

Chaos, or entropy, in the enterprise is a success-killer: It angers and enervates those who buy from, invest in, write about, partner with, supply, and work at your company.

Entropy drives patrons to go elsewhere and employees to build fiefdoms, go rogue, sabotage, and battle each other in the dreaded internecine wars—all of which decimate your top and bottom lines. In other words, entropy is a brand-killer.

Entroprises make bad hiring decisions, resist firing incompetent employees, implement poor policies, are slothful, develop off-target or inferior products, and are impossible to fathom and distinguish from their competitors.

Are you a chief *entroprise* officer?

The acorn of the entroprise: the nonexistent, nebulous, or horrible brand. When employees get no direction, or confusing or bad direction, they respond chaotically, in ways that *they* choose, ways that harm your business.

That said, sometimes the brand *is* crystal-clear, but the navigation system or the navigators betray it. Apple Maps, for example, has taken me and millions of others to incorrect locations and deadends, because of faulty algorithms or data.

Likewise, inexperienced marketing teams make false assumptions and use faulty logic—putting their companies' strong brands in jeopardy. This can happen when firms grow rapidly and hire new employees faster than execs can properly vet, train, and task them.

Example: Impulsively using ineffective social media as branding and selling megaphones, swallowing the false hype that these conduits have supplanted traditional megaphones.

Often, CEOs don't know about problems like this until they fester. That's why knowing your company is paramount.

Other times, the CEO injects false assumptions and faulty logic of his own into the mix.

On March 4, 2016, *Fortune's* Jennifer Reingold wrote a scathing piece about Zappos, the online shoe retailer owned by Amazon. Once considered a "Best Company to Work For," Zappos had fallen from grace. Why?

Tony Hsieh, the CEO, replaced traditional hierarchical management with self-management: a *holocracy.* Nonsense.

Newsflash: People can't manage themselves—*they just can't.* Perhaps you can guess what happened next.

After one year, almost one-third of Zappos's employees had bolted. Morale sank. And, those who remained responded negatively to *Fortune's* destiny-centric survey question: *Do you think management has "a clear view of where the organization is going and how to get there"?* Shocking!

Imagine any high-level sports team winning one game without an authoritative, order-spewing coach. You cannot. It is impossible. Evidently, Zappos imagined the undoable.

That Vision Thing

Many accused George H. W. Bush of lacking vision—one of the reasons he was a one-term president. In fact, when pressed about it, he retorted, "Oh, that vision thing."

Vision begets purpose and direction.

In 2017, Comparably, a firm that studies corporate culture, found that 37 percent (almost one in four) of 10,000 tech-firm employees said their CEOs lack vision and strategy, and need to improve themselves.

This percentage should not exceed one percent!

Employees may not be branding experts but know when their companies lack purpose and direction—*vision*—and know that their CEOs are responsible for this deficiency.

This lack of vision exists, especially in tech precincts, in companies that are product/technology-centric.

Enterprises, regardless of age, size, industry, customer category, or geography, should revolve around their brands—*never their products and technologies.*

Every employee, in every department, at every level, should be able to recite his company's brand—its customer-validated value proposition. So should every board member, investor, channel partner, and customer. Verify it or fix it.

CHAPTER EIGHT

Entroprise Profiles

What Is Xerox?

On March 24, 2014, in Manhattan, Andrew Nusca of *Fortune* interviewed Ursula Burns, then-CEO of Xerox, before her company's Simple@Work conference. He asked about the perception of Xerox and her biggest challenges of that year.

Below are six excerpted paragraphs [bold added for emphasis] of Burns responding to Nusca:

*One of the things Xerox can help do—and we're not the people who do it all—but **we can allow more goodness to be spread around the world, easily.** And that's what, fundamentally, our business is about.*

*The first question they asked us was, after they put in our technology, "I have your technology over here, and I have some other guy's technology over there. All of it is managing printing and copying and faxing. Can't you just manage it all for us?" That was literally our first foray into services. **We started to realize that our skillset of using technology to solve problems—our*** *brand—gives us the trust we need to do this.* So we said, "Okay, we can do that." And we went back to the lab to invent it. So this lack of fear around the idea that our smart people can solve this problem actually kept dragging us toward using the assets that we had in a broader way.

We have a great brand—people don't just know our name; with it, they associate great things— around innovation, around being global, around— and this is the most important thing—staying with customers. We're not perfect, but we're not going to leave them hanging. We'll work with them. We'll make sure that they actually shine, even if it means some skin off of our hide. So these assets that we have —people that are amazing, and have been at our company from their first job, and not because they're stranded there, but because we have that team perspective—those assets should be able to be used close to our core business in more ways.

Because we have such a strong brand that is associated with something historic, the thing that people most misunderstand about Xerox is what we actually do today. If you ride around New York City and you get a ticket from one of those little flashing things in the road, or if you have an E-ZPass transponder, or if you live in the state of California or Indiana or Texas or Florida or use public transit in

Philly—guess who is behind all of the stuff you see? That's Xerox. So we're the people behind all that. **If you fly on a plane, and it comes from the biggest plane-maker in the world, the documentation for that plane and the tracking of the configuration of that plane—we do it. People don't know that.**

The business community is starting to understand more and more. We can actually have an effect on the business community—that's one of the reasons we have an event like this. We have over 800 customers here. But consumer knowledge is something that we're going to have to be pretty careful about. **It's important to have consumer knowledge because we employ people every day—because we have pride in our company and in people knowing what the hell we do for a living. We're going to have to figure out a way to really supercharge our presence in the consumer's mind based on what we actually do.**

The best correlation I can draw is Intel. They're doing all these chips; nobody knows about this company. Advanced Micro Devices—who the heck knows what they do? ADM [Archer Daniels Midland Co.]—you don't know them! You know peanut butter, but you don't know ADM. **So we're going to have to put out, over time, a more consumer-facing set of communications. But that's number two, not number one. Before we do that, we have to knock** the ball out of the park on business-facing communications. We're getting closer every single day. We do commercials and that stuff, and that's really important. Now our commercials are about our customers—we don't do commercials about Xerox. The best advocates we have are our customers. It's a really small community, and they all kind of herd in their like tribes. So if you can do something really good in transportation in Alabama, you can actually probably sell transportation in Mississippi—unless it was a Republican-Democrat thing, that kind of ridiculousness. The best communicators of what we can do are other businesspeople; our customers talking to other customers. It really is amazing.

After we get that pretty well settled down, we're going to have to really think of our branding. Not the brand name, but our messaging for regular consumers, and have a little bit more consistent and pointed communication to them. Because I do want our employees to be proud. I want people to know that we do this. We have to hire people all the time, and I don't want them to say, "Xerox? What the hell do they do?" I want them to say, "Yeah, these guys do a lot on analytics and a lot on ..." Because they understand. It's not for the buying in the short term that we need consumers *to understand. It's for the respect and longevity of the company.* [Emphasis added.]

What is Xerox? After reading Burns's words, I couldn't begin to explain its brand—because *she* can't. Can you? She's all over the map. *More goodness to be spread around the world, easily?* Seriously? Imagine being a Xerox employee, responsible for innovating or selling, with the aforementioned as the purpose and direction.

Murkiness has consequences. In January 2016, famed investor Carl Icahn forced brandless, undervalued Xerox to split in two. Ironically, Ursula Burns had purchased Dallas–based Affiliated Computer Services (ACS) for $6 billion in 2010, to mask her company's weak brand. Because of Icahn, she had to spin off ACS into a new company called, drum-roll, please, Conduent. Xerox was back to copiers and printing.

In November 2019, under CEO John Visentin, Xerox began a five-month quest to subsume HP, more than 3X its size. Why? Paper-based printing is waning. Xerox doesn't know what to do next. An acquisition is a common panacea for lack of direction. Which customers would benefit from this merger? Nobody knows. But, Carl Icahn, who owns 11 percent and four percent of Xerox and HP, respectively, was in favor of the merger. Because of the COVID-19 pandemic, HP called off the merger in late March 2020.

Murkiness is a common problem in enterprises around the world. It begets the weak keystone of the entroprise and underscores the importance of intrabranding.

Theranos's Tyranny

The antithesis of poor communication is militantly and tyrannically coercing all employees to believe and repeat a lie. This technique works exceeding well, until it doesn't.

Enter Theranos, phony blood-testing startup founded by Elizabeth Holmes, whom the DOJ has charged with felony wire-fraud and conspiracy for deceiving patients, doctors, retail chains, investors, and government agencies.

Eventually, petrified employees ratted on Theranos to the *Wall Street Journal*, the FDA, and other entities.

Read John Carreyrou's *Bad Blood: Secrets and Lies in a Silicon Valley Startup* and watch HBO's *The Inventor: Out for Blood in Silicon Valley,* both of which will leave you astounded and incredulous about how easily Holmes and former COO Ramesh "Sunny" Balwani duped so many people—including Henry Kissinger, Rupert Murdoch, George Schultz, David Boies, Betsy DeVos, and General James Mattis—into believing that she had produced a viable desktop machine that could perform over 200 assay tests from a single drop of blood.

Why was Elizabeth Holmes able to defraud an A-list of sophisticated and normally savvy luminaries?

First, she's a woman. They *wanted* to believe her. In today's women-are-a-priority culture, they rushed to back a "female Steve Jobs." And, craving more notoriety, they bought Holmes's "my invention will save the world" mantra.

The second reason, which you *must* copy in your company, Holmes exhibited deep *passion and conviction.*

Without passion and conviction, you can't sell anything to anybody, internally or externally.

The WeWork Sham

Enter WeWork, a spreadsheet pipedream.

WeWork, which lost $1.9 billion in 2018, proclaimed to provide ethereal office space for enlightened Millennials and Gen-Zers. Its business model was to rent office space (100 cities in 29 countries), typically in a skyscraper, then convert said space into a utopian dream, then re-rent it to woke entrepreneurs.

The company's promise in 4Q19:

"Give our members flexible access to beautiful spaces, a culture of inclusivity, and the energy of an inspired community, all connected by our extensive technology infrastructure. We believe our company has the power to elevate how people work, live, and grow."

Reality: An entrepreneur should rent the cheapest space he can find and scrimp, scrape, and survive on spartan amenities until the company becomes viable and profitable. Comfort then grows with the bottomline.

But, that rarely happens in the modern, cushy world. Millennials and Gen-Zers, many of whom still live with their parents, bristle at the hardships of the real world—and, therefore, are rarely forced to experience them.

What was WeWork's brand? An extension of Mom & Dad's home. It was not a business. That's why an unproven entrepreneur cannot afford WeWork's utopian dream.

WeWork's IPO, consequently, failed to materialize: the company's value tanked in a few weeks from about $50 billion (based on what?) to $20 billion (based on what?).

Masayoshi Son, CEO of Softbank, WeWork's largest investor, forced Adam Neumann, WeWork's embattled CEO, to leave—because of his alleged eccentric behavior and drug use—and paid him $1.7 billion to do so.

The question is, why did WeWork's investors awaken so late in the game? Neumann's weirdness wasn't news.

Answer: Despite all the investors' woke fantagasms, Neumann didn't and couldn't deliver *profits*.

WeWork's so-called business model, which violates all economic principles, was exposed publicly because of the delayed IPO. That model can't work and was obviously flawed when investors, including SoftBank, funded this mirage.

Why did they fund it?

The inexplicable, pathological, universal worship of Millennials and Gen-Zers—a worldwide disease, invented in America. The gullible investors, otherwise savvy, evidently believe that whatever these overgrown tikes do and say is golden.

In May 2020, Neumann sued SoftBank for abandoning its proposed $3 billion rescue of WeWork, one month before.

Management in today's infantilized culture is hardly management. If Google's execs will terminate a $250-million

JEDI project for the US military, because its Millennial Mob employees didn't like it, yet perform the same kind of work for the Chinese military, then management as we know it is dead.

Perhaps the so-called adults should revisit what it means to be an adult.

Managing people the way they want to be managed is the epitome of silliness: it can render managers useless and kill a company. The idea is a politically correct hoax, and corporate executives must stop uttering it.

Millennial worship hurts companies by leading them down unprofitable rabbit holes. And, it especially hurts Millennials—because they'll never grow up. Then what? Who will raise and manage the next generation? Robots?

Boeing's Big Blunder

It was bad enough that Boeing was poorly managed under Dennis Muilenburg, the ousted CEO. Then, COVID-19 hit, and the company slowed to a halt: it delivered only six planes and four planes, in April and May of 2020, respectively, its lowest output in 60 years.

I opined about Boeing's colossal branding blunder in my *Newsmax* column. Below, I've excerpted and drawn from that column, revised it as needed, and amended it with new information that surfaced after its publication.

Responding to the ongoing safety crisis over the grounded Boeing 737 Max 8 and Max 9 planes, at the Aspen Ideas Festival in June 2019, former CEO Dennis Muilenburg inartfully said, "I don't see a need to change the name of the airplane. To me, this is not a marketing or branding exercise. I know that's important—certainly it impacts the public view—but the most important thing is safety."

With those unfortunate words, Muilenburg proved that he knows and cares little about branding. Moreover, safety wasn't at all important to him.

The distrust of Boeing, once a $101-billion behemoth, began when two of its 737 Max airplanes crashed and killed 346 passengers, in Indonesia (October 2018) and Ethiopia (March 2019), respectively. Panic spread so quickly after the second crash that President Trump, echoing other countries, ordered the complete grounding of these jets.

This Boeing debacle has so damaged the finances of worldwide airlines, which had ordered or were already flying the 737 Max planes, that Kevin McAllister, the eventually fired head of Boeing's commercial-aircraft division, apologized at the June 2019 Paris Air Show for the crashes and his company's failures.

The culprit of the two crashes was a malfunctioning MCAS (Maneuvering Characteristics Augmentation System), Boeing's anti-stall system. Designers moved the big engines on the new 737 Max farther forward and higher on the wings, making it more likely to stall. Hence, the need for MCAS.

Amazingly, many 737 Max pilots either didn't know their planes were equipped with MCAS or had received inadequate or no training on it from Boeing—impossible in a company that values communication.

Moreover, **Boeing's engineers**, one year before the two crashes caused 346 deaths, *were aware of and concealed the flaws of MCAS.* Safety, apparently, was not "the most important thing" anywhere at Boeing.

Worse, because of its cozy relationship with Boeing, the FAA (Federal Aviation Administration) had certified the 737 Max planes with perfunctory scrutiny.

In April 2020, the FAA mandated that Boeing repair two more software faults: one in a flight-control computer that could lead to a loss of control from the horizontal stabilizer; the second that could lead the autopilot feature to disengage during final approach.

Despite all that, the FAA announced in May 2020, one month later, that it would *continue allowing Boeing to play a major role in certifying its own planes.*

It's reasonable to conclude that, had President Trump not joined other countries in grounding the 737 Max, in March 2019, hundreds more people would have died. As it is, Boeing agreed, as of November 2019, to settle at least 63 lawsuits from families of the crash victims.

Back to Muilenburg's dismissive, stupid comment about branding and renaming the 737 Max. Ordinarily, a name—product or company—is *not* a brand. A brand is intangible. It's the emotional connection customers have with a given supplier. Example: Apple's customers, in years past, camped out overnight to get its newest product the next day. *That* is strong emotional connection to a supplier.

NOTE: Exception to the brand definition above: *when a name becomes so unredeemably toxic, so hated and reviled, that it turns into a permanent liability.*

By opining that safety, not marketing or branding, is most important to him, Muilenburg unwittingly put Boeing's brand into position for a perilous stall—because *safety and branding are interdependent.*

Before boarding, passengers implicitly trust that their airplane is safe, based on presumed and demonstrated safety. **Trust is, therefore, a huge part of every supplier's brand.**

For Muilenburg to cavalierly demote branding while jeopardizing his company's brand is a sign of pure hubris.

Source of hubris? Record 2018 profits ($10 billion in net earnings), a backlog of 5,900 planes (end of 2018), and only one competitor: Airbus, headquartered in Holland.

Hubris begets complacency and poor communication, which beget engineering blunders, shortcuts, product defects, and cover-ups. **This is why intrabranding is so critical.**

Not realizing the full extent of Boeing's problems, airlines hungered for the 737 Max. At the aforementioned Paris Air Show, International Consolidated Airlines, parent of British Airways, provisionally ordered 200 of these planes from Boeing—contingent on the FAA's certification, which was in doubt.

To compensate for its previous hands-off approach to Boeing, the FAA began putting the 737 Max though the wringer and, in late June 2019, announced another MCAS problem. Consequently, nervous shareholders sent Boeing's stock downward almost three percent.

If the 737 Max is unfixable, Boeing's brand will become untrustworthy. A stalled reputation is infectious and difficult to right. Everyone is asking, how many other MCAS-like situations exist at this aerospace company?

Had branding been Dennis Muilenburg's top priority all along—*safety, comfort, economy, speed*—and **ingrained corporatewide**, engineers wouldn't and couldn't have hidden MCAS's flaws. And, today, trained pilots would be flying the 737 Max, not reading about its possible demise.

By January 2020, Boeing's picture had dimmed. The 737 MAX saga had lingered too long, and I averred that Boeing had to jettison its MAX nameplate.

To wit: Steven Udvar-Hazy, chairman of aircraft-leasing firm Air Lease, which at the time had 150 of Boeing's grounded 737 MAX jets on order, urged Boeing to drop the "damaged" MAX label—to avoid further destroying the plane's value. At the Airline Economics aviation-finance conference, in Dublin, Ireland, Udvar-Hazy admitted:

> "We've asked Boeing to get rid of that word MAX. I think that word MAX should go down in the history books as a bad name for an aircraft."

Marc Rudov: *Once customers publicly hate your product's name, it's dead. Bury it.*

It got worse. Internal emails and text messages from employees, delivered to FAA and congressional investigators, revealed a culture that blatantly disregarded human life:

"This airplane is designed by clowns, who are in turn supervised by monkeys," one Boeing employee wrote in a 2017 instant message exchange apparently bashing fellow colleagues at the company.

"Would you put your family on a Max simulator trained aircraft? I wouldn't," another employee asked a coworker in a 2018 conversation before the first crash. "No," the person responded.

"I still haven't been forgiven by God for the covering up I did last year," one employee wrote in 2018, referencing interactions with the FAA.

Intrabranding 101: Culture emanates from the CEO and C-suite. A branding-oriented company puts customers *first*. A revenue-oriented company like Boeing does not.

Boeing suffered multiple losses in 2019 and 2020:

- Stock dropped from $440.62 on March 1, 2019, to $306 on January 21, 2020, to $166 on July 29, 2020.
- July 2019: recorded its largest quarterly loss ($2.9B) in its history
- FY 2019: annual revenues dropped to $76 billion ($101 billion in 2018); annual loss of $636 million (2018 profit of $10.5 billion).
- July 2020: first half of 2020, Boeing posted a loss of $3.04 billion, or $5.31 a share. Revenue dropped 26 percent to $28.7 billion.

Boeing's board fired Dennis Muilenburg, in December 2019, and replaced him as CEO with Dave Calhoun, who was serving as the company's board chairman.

For his colossal failure, Muilenburg left Boeing, where he'd worked his whole career, with a $62-million parting gift.

Treasury Secretary Steven Mnuchin predicted that the Boeing debacle, because of its effect the airline industry, could knock a half-percent off America's GDP growth.

In January 2020, America's largest aircraft builder was seeking to borrow $10 billion, to compensate for its lost revenues. In May 2020, instead of accepting a COVID-19 bailout, it borrowed $25 billion through a bond offering.

There's a huge cost to subordinating branding: failure. Boeing is paying it now.

Engineers hid MCAS flaws *because doing so was consonant with the culture*. Because business for Boeing was so easy for so long, the company rested on its laurels, became lazy and complacent, and managed by assumption.

Mr. Calhoun, the new leader, has much to do to convert the entroprise to an enterprise and instill a new culture there. As of June 2020, CEO Calhoun telegraphed no plan to ditch the 737 MAX name.

Yet, on July 2, 2020, Calhoun proved himself just another woke, virtue-signaling, cancel-culture CEO. Under the guise of "diversity & inclusivity," and with no sense of irony, he forced his SVP of communications, Niel Golightly, to resign after being on the job only six months. Why? *One employee's complaint*. Golightly's sin? In 1987, 33 years ago, the 29-year-old Navy pilot penned an article in which he opposed women fighting in combat. Even though Golightly has since changed his mind and has fallen on his sword in remorse, no matter. He's now a nonhuman.

Say or do *anything,* at *any* point in your life, that counters the "progressive agenda," spineless CEOs will deem you irredeemably shameful, proving, hypocritically, that *there is neither diversity nor inclusivity in corporate precincts.*

Ford's $250-Million Edsel Disaster

On November 19, 1959, after five years of committee-based planning, designing, tooling, producing, overpromoting, and selling its much-heralded new car—at a whopping cost of $250 million—Robert McNamara, president of Ford Motor, canceled the Edsel, the name now synonymous with flop.

The hubristic, starry-eyed dreamers at Ford forecast sales of 400K Edsels per year. In fact, *total* Edsel sales were around 116K, less than half the projected break-even point. Ford built only 2,846 Edsels—including 76 convertibles—in 1960, the final year. The last model was merely a mildly restyled Ford with a conventional grille.

With the Edsel, named for Henry Ford II's father, Ford Motor simply did everything wrong: most notably focusing on besting competitors, not serving *customers*; making a product too similar to its other cars; and assuming, *with no testing,* that it would be a hit. Ford Motor was a true entroprise, if there ever was one, with no concept of intrabranding.

When all other carmakers had a few models per car, Edsel featured 18 on launch day, September 4, 1957, during a recession, when cheaper, more-compact cars were in vogue.

On October 13, 1957, Ford funded a one-hour, highly rated variety program on CBS called *The Edsel Show*, which starred Frank Sinatra, Bing Crosby, Rosemary Clooney, Louis Armstrong, and Bob Hope. The car was featured on the show. Couldn't miss, right? Wrong. Nobody cared about the car.

Americans ridiculed Edsel's vertical, front-grille design, thinking it resembled a toilet seat, a vagina, a horse collar, an Oldsmobile sucking a lemon. The rear turn signals, shaped like reverse-pointing arrows, confused other drivers. Owners mistook the gear-selecting pushbuttons on the hub of the steering wheel for the horn. The list of flaws goes on and on.

The thrill starts with the grille

(And never seems to end)

EDSEL

On the day the Edsel died, *TIME* magazine wrote:

The Edsel was a classic case of the wrong car for the wrong market at the wrong time. It was also a prime example of the limitations of market research, with its "depth interviews" and "motivational" mumbo-jumbo.

On the research, Ford had an airtight case for a new medium-priced car to compete with Chrysler's Dodge and DeSoto, General Motors' Pontiac, Oldsmobile, and Buick. Studies showed that, by 1965, half of all US families would be in the $5,000-and-up bracket, would be buying more cars in the medium-priced field, which already had 60% of the market. Edsel could sell up to 400,000 cars a year.

After the decision was made in 1955, Ford ran more studies to make sure the new car had precisely the right "personality." Research showed that Mercury buyers were generally young and hot-rod-inclined, while Pontiac, Dodge, and Buick appealed to middle-aged people. Edsel was to strike a happy medium. As one researcher said, it would be "the smart car for the younger executive or professional family on its way up."

To get this image across, Ford even went to the trouble of putting out a 60-page memo on the procedural steps in the selection of an advertising agency, turned down 19 applicants before choosing Manhattan's Foote, Cone

& Belding. Total cost of research, design, tooling, expansion of production facilities: $250 million.

The flaw in all the research was that, by 1957, when Edsel appeared, the bloom was gone from the medium-priced field, and a new boom was starting in the compact field, an area the Edsel research had overlooked completely.

When you hear people blaming a product failure on the inadequacy of market research, assume that the source of this criticism knows nothing about market research (read Chapter 2, Market Review).

Market research is *customer* research, because the market is *customers*, and has nothing to do with products, and it contains no "motivational mumbo-jumbo." Product research, in contrast, which most people wrongly think is market research, is limited to the favorability of features and functions of an existing or future product.

Ford brought in David Wallace from *TIME* magazine as manager of market research. Few insiders listened to him. He was responsible for picking a name for this new car. The first name was Edsel, but The Deuce (CEO Henry Ford II) didn't like it.

Wallace hired ad agency Foote, Cone & Belding (FCB) to find a name. FCB narrowed a list of 6,000 entries to the same four that Ford insiders had chosen: Citation, Corsair, Pacer, and Ranger.

Frustrated, Mr. Wallace enlisted the talents of Pulitzer Prize-winning poet Marianne Moore to help pick a name. Her most-insane suggestion: Utopian Turtletop. Seriously. *Never brand by committee. Never.*

Finally, Ernest Breech, board chairman, subbing for the vacationing Henry Ford II, picked ... drumroll ... Edsel.

Angus MacKenzie, writer at *Motor Trend*, explained in July 2009 why Robert McNamara canned the Edsel:

> *McNamara was one of a group of young officers from the US Army Air Force's Office of Statistical Control, hired by 28-year-old Henry Ford II in 1946 to help rescue the ailing automaker. The "Whiz Kids" helped install fiscal and process discipline at Ford, the management of which had become ever more ad hoc as aging founder Henry Ford's dementia grew more apparent. By 1948, McNamara had assumed the role of leader of the Whiz Kids and was clearly on a trajectory to the top. By 1955, he was general manager of Ford Division [and became president in 1960].*

> *McNamara showed his iconoclastic product streak early by authorizing a four-seat Thunderbird, much to the horror of purists who saw the original two-seat T-bird as a potential rival to Chevrolet's Corvette. He was implacably opposed to the Edsel program, arguing from the outset that, if Ford needed to move into the mid-price market, it would be better to simply upgrade the*

top-of-the-line Ford than waste money creating a new car, a new division, and a new dealer network.

McNamara was right in both cases.

First-year sales of the four-seat Thunderbird exceeded total sales of the two-seater since launch. And, Edsel, part of an ambitious plan to tackle GM's Buick, Oldsmobile, and Pontiac divisions that involved building three basic bodies (small, medium, large), across five divisions (Ford, Edsel, Mercury, Lincoln, Continental), foundered in the teeth of the Eisenhower recession. With a little help from politically adroit McNamara.

Sources inside Ford insist McNamara effectively killed the Edsel before the first car had even been sold, deliberately letting slip at its launch [that] the car would be discontinued. The day after the first Edsel went on sale in 1957, McNamara was made a group vice-president responsible for all FoMoCo cars and trucks, and, sure enough, he began hacking away the division's budget almost immediately. Within months, he had reduced Edsel's future product plans to little more than a different grille for 1960 Fords.

Scores of other articles, books, videos, and business cases exist to explain Edsel's demise. Bill Gates claims on his blog that *Business Adventures,* by the late John Brooks, which Warren Buffet gave him, is "the best business book I've ever read." He particularly likes its Edsel chapter:

In "The Fate of the Edsel," he [Brooks] refutes the popular explanations for why Ford's flagship car was such a historic flop. It wasn't because the car was overly poll-tested; it was because Ford's executives only pretended to be acting on what the polls said. "Although the Edsel was supposed to be advertised, and otherwise promoted, strictly on the basis of preferences expressed in polls, some old-fashioned snake-oil selling methods, intuitive rather than scientific, crept in." It certainly didn't help that the first Edsels "were delivered with oil leaks, sticking hoods, trunks that wouldn't open, and pushbuttons that...couldn't be budged with a hammer."

With a pricetag starting at $2500 and increasing to $3800, each Edsel *lost* $3200. $3200! Yet, the arrogance at Ford, *common in entroprises*, was beyond the pale. J.C. Doyle, an Edsel marketing manager, blamed *customers* for his company's fiasco. He told author John Brooks: "What they'd been buying for several years encouraged the industry to build exactly this kind of car. We gave it to them, and they wouldn't take it. Well, they shouldn't have acted like that. And, now the public wants these little beetles. I don't get it!"

Ford's abysmal failure validates intrabranding. Instead of beating its competitors, Ford beat itself.

CHAPTER NINE

Innovation

Every year, *Fast Company* and Boston Consulting Group (BCG) rank their "most innovative companies." BCG decided that, in 2020, Apple had surpassed Alphabet (Google) as the most-innovative company. Interestingly, Huawei, the Chinese Communist Party-controlled telecom-equipment company, jumped 42 notches to sixth place from 2019.

How do *Fast Company* and BCG determine their rankings? And, how do they define innovation and innovative?

Great question.

Marc Rudov's branding philosophy: *If one can't articulate it, one can't execute it.*

In its 22-page report, BCG states that "an innovation culture is notoriously hard to describe or assess." This is unacceptable. BCG describes its methodology thusly:

> The BCG most innovative companies ranking is based in large part on a survey of 2,500 global innovation executives (63% C level, 37% senior vice-president or vice-president level) who were polled from August 2019 through October 2019. We assess companies'

119

performance on four dimensions and then take an average of normalized scores to calculate the overall ranking. This year, as noted in the text, we added a new scoring dimension that captures each company's variety and intensity of boundary breaking, by assessing its ability to breach established industry entry barriers and play in an array of markets [industries] outside its own. These four dimensions are:

- Global "Mindshare": Number of **votes** received from all global innovation executives
- Industry Peer View: Number of **votes** received from executives in a company's own industry
- Industry Disruption: Herfindahl-Hirschman Diversity Index of **votes** across industries
- Value Creation: TSR [total shareholder return] including share buybacks from January 2017 through December 2019 (three years).

Three out of four of BCG's scoring dimensions above are "votes" (impressions) from execs in other companies—hardly helpful. TSR is calculable, but we don't know in these rankings its root-cause (efficiency, low overhead, lucky product, share buyback, etc.).

BCG never explicitly, clearly, succinctly defines innovation and innovative—a universal problem. Again, *if one can't articulate it, one can't execute it.*

The Boston-based consultancy does, throughout its 2020 report, drop hints, breadcrumbs, and components of an innovation definition. Here are nine:

- Make success sustainable and repeatable
- Adapt to rapidly shifting patterns of supply, consumer demand, and commerce methods
- Conquer an unrelated industry (e.g., Amazon in healthcare). **WARNING: most companies can't and shouldn't try this, as the era of failed conglomeration proved.**
- Establish clear, shared understanding of what innovation means [**intrabranding**]
- Unite the whole org behind the innovation strategy [**intrabranding**]
- Garner a large portion of sales from products or services ≤3 years old [**not necessarily valid**]
- Develop a systematic approach to innovation
- Tie new ideas to customer needs [**branding**]
- CEO must drive innovation [**intrabranding**].

Fast Company, the tech-oriented business publication, explains its ranking methodology this way:

> *This year,* Fast Company's *editors and writers sought out groundbreaking businesses across 44 sectors, including every region of the world. We also judged nominations received through our application process. We assess each company on a combination of innovation and impact, with a focus on what it's*

accomplished in the past year. The 434 organizations we honor here lead their fields and are transforming the world.

Our annual ranking of the businesses making the most profound impact on both industry and culture showcases a variety of ways to thrive in today's volatile world. Snap's work building out an augmented-reality platform to the delight of its universe of users, content creators, and advertisers means that we're all able to experience the future of computing being built before our eyes today. Shopify is one of several companies on this year's list powering a larger ecosystem of creators to help them make money online. Sustainability is another significant trend in which we're seeing meaningful innovation: The troika of CaaStle, ThredUp, and Trove are helping to build out a circular economy in the fashion business, and Footprint is the quiet giant in the sustainable packaging space, reinventing everything from the microwaveable meal bowl to the fast-food restaurant coffee cup and helping consumer packaged goods transition from single-use plastic to a biodegradable alternative.

Learn how these and 44 other companies are creating the future today, plus see our top-10 lists of the Most Innovative Companies by category, from advertising to wellness.

Fast Company did not define innovation, despite invoking it twice. "Groundbreaking" doesn't necessarily mean innovative, nor does "transforming the world."

Innovation Ain't Disruption

The more you explore terminology, the more you'll realize that innovation, like brand, market, and entrepreneur, has many, nebulous "definitions"—most of them wrong.

No one can achieve anything without a clearly defined quest. That's why enterprises have such difficulty innovating.

Putative experts have written volumes of scholarly and not-so-scholarly books and articles on innovation—some so complicated that they muddy the waters.

Some equate innovation to disruption. Wrong. There's an obsession in tech circles to disrupt everything. Who likes disruption? Other than VCs (venture capitalists), divorce lawyers, and leftist rioters/looters, nobody.

DISRUPTION
CUSTOMERS HATE IT
MarcRudov.com

Customers hate disruption. Divorce is disruption. Rioting is disruption. A pandemic lockdown is disruption. Disruption is painful and undesirable.

In general, people prefer and embrace *gradual* change, which is manageable. The key to successful innovation is diligently maintaining a steady rate of *incremental* change.

Disruption is the bane of innovation: it focuses on *supply*, one-upmanship against competitors, and industry dynamics—*and not on demand (customers)*.

The Dilbert strip below illustrates this perfectly.

Enhancing Customers' Lives

Innovate is derived from the Latin *innovare*: to make new. **Newsflash: new isn't necessarily innovative. Nor is revolutionary.** To wit: 95 percent of patents—inventions—are sitting on the shelf, unlicensed and uncommercialized.

Put another way: *A new and revolutionary product isn't necessarily an innovation.* Just review the Edsel saga in the previous section if you doubt that.

NOTE: Unless the "innovator" has improved and enhanced the lives of customers, *according to them,* there is no innovation. *Never confuse invention with innovation.*

Moreover, without a formalized means to assess whether customers' lives are enhanced (excluding bogus social media), innovation is impossible.

NOTE: This axiom applies regardless of company size, age, or location, as well as industry and customer category (military, industrial, scientific, commercial, retail, financial, consumer, etc.). **Violate it at your peril.**

Take the use of artificial intelligence (AI) in customer service. Many in Corporate America and the business media blindly and blithely accept AI as a great innovation—because it's in vogue and allegedly saves money by shortening process time and replacing humans.

But, when an impersonal, detrimental, alienating experience induces the customer to yell at store clerks and telephone techs, waste his time, despise the supplier, and abruptly hang up the phone (read my article "Best Buy? You Cannot Be Sirius!") on clueless idiots, AI is not an innovation.

Amazon's Alexa is a well-known consumer-based example of AI. By issuing voice commands to Alexa, users query the temperature, request music selections, set reminder alarms, and switch on and off their appliances.

Is Alexa improving and enhancing their lives?

Despite becoming lazy, helpless addicts, the majority of Alexa users likely would answer in the affirmative—even though Amazon records (spies on) their conversations and sexual activities. Oddly, upon learning about the spying, most owners express temporary shock, shrug off their dismay, and continue their habitual addiction.

Count me out: I won't own an Alexa product, because it cannot improve my life, as I define it. To each his own.

Continuous Improvement

Continuous improvement (CI) is a critical business process, based on neverending *incremental change*. Not huge change. Incremental change.

CI lies at the core of widely deployed management paradigms such as Six Sigma and Toyota's Kaizen (lean manufacturing)—*and is in the DNA of innovation*. Customer-driven innovation.

Why?

It works. All success stems from incremental change. Sales training, marriage counseling, corporate restructuring, raising and educating children, and building customer loyalty, among other tasks, depend on *gradual* change.

Sadly, most people are procrastinators: they wait until things are severely broken to act. Then, they are forced to make big, expensive, slow, unpopular changes.

Remember: Put a frog in boiling water; it jumps out. Put it in cold water and gradually turn up the heat; it cooks.

That's why there's a false perception that innovation is about making big change, for change's sake. Wrong.

Innovation = improvement, *as customers define improvement,* **not invention.**

It's also easier to communicate incremental changes, thereby affecting the success of intrabranding.

Accordingly, consistently innovative companies excel at intrabranding, which is based on *superior corporatewide communication.* Failure in intrabranding is the CEO's fault.

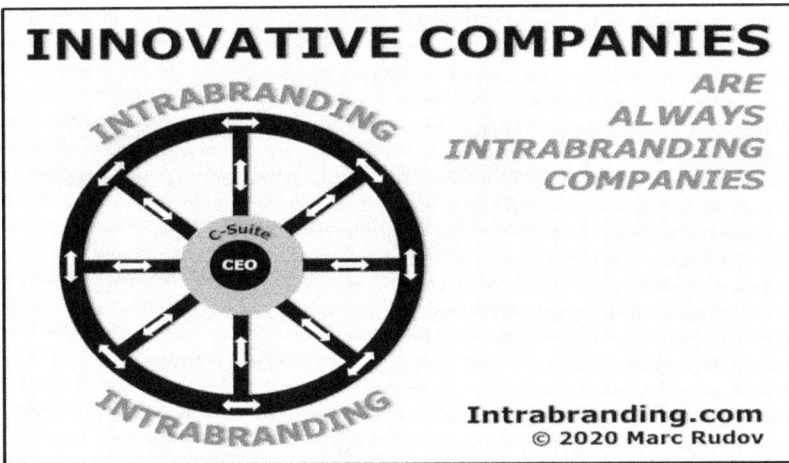

Four Pillars of Continuous Improvement:

1. Incremental changes (cheaper/easier to implement *and adopt* than large ones)
2. Employee participation/contribution (closest to the action)
3. Constant communication/feedback (intrabranding)
4. Measurable & repeatable.

People hate and resist big changes, and companies flop when implementing big changes. So, always make a continuous stream of incremental changes. All changes must be measurable and repeatable, and must be communicated clearly—to and from all corners of your company.

The Branding-Innovation Cycle

Absent an internally understood, endorsed, and deployed brand—*which establishes a company's purpose and direction*—no enterprise of any size will excel at innovating.

The objective of intrabranding, therefore, is to get employees to understand, endorse, and deploy the brand: use it to design (and/or redesign) and sell products and services.

Entroprises do *not* have that objective.

When I'm dealing with execs and employees of almost every firm (especially those in customer service), it becomes obvious, quickly, that *the left and right hands have never met.*

When innovating, the fundamental question should be, *what new/tweaked offering should we produce, for whom, and why?* And, which internal group should lead this effort?

Your brand *must* answer this question.

Without knowing what your customers want and need, and how they feel about your company—and lacking the concomitant purpose and direction that a strong brand confers—you can neither hatch nor improve an offering (product or service) with success.

As the Edsel debacle proved, branding by committee doesn't work, hasn't ever worked, and never will work. But, the small, focused group tasked with innovating, or achieving any corporate objective, *must* be on the same page, the right page, and speak the same language—clearly and effectively.

After crafting a strong brand, successful innovation depends on cohesive cross-discipline execution.

Hence, the Integrated Innovation Institute at Carnegie Mellon University (CMU) in Pittsburgh unites the disciplines of engineering, [industrial] design, and business to teach its students a holistic approach to building *impactful solutions that create value for customers.* CMU is spot-on: the inability to direct all your stakeholders—including distributors and franchisees—will stymie or nullify your innovative efforts.

Now, assuming your company's product or service is unique (and it should be), it *may* behoove you to patent and/or trademark it. Tread carefully, because not all products are economically worthy of patenting or trademarking.

Finally, you must be able to verify that your company's innovation is, in fact, an innovation: *it enhances the personal or business lives of your customers,* **according to them.**

BRANDING-INNOVATION CYCLE

VALIDATE/CREATE BRAND

DOES CURRENT OFFERING MEET BRAND'S PROMISE?

TWEAK/CREATE OFFERING TO MEET BRAND'S PROMISE

PATENT AND/OR TRADEMARK OFFERING, IF BENEFICIAL

VERIFY INNOVATION'S SUCCESS

© 2020 Marc Rudov

Intrabranding.com

As the figure above depicts, innovation is a continuous process, dependent on a strong brand for clear guidance.

Without said clear purpose and direction—*understood, endorsed, and employed by all employees at all levels*—what, how, and why are you innovating?

Building a strong brand means knowing and revering customers—and communicating to them in their language.

Per Evan Facher, PhD, director of the University of Pittsburgh's Innovation Institute, the university's central hub for research-commercialization: "Technology for technology's sake doesn't help them [customers and investors]. It has to address a pain that they have."

Never assume that, to innovate, your company must crank out a new product or an iteration of an existing product. That may or may not be true. Sometimes, all that's necessary is a small, tactical or attitudinal improvement, involving little or no technology.

Why do companies insist on putting impossible-to-remove price stickers on their products? It creates work for customers, requiring a half-hour of rubbing Goo-Gone on the sticker until it comes off. Insane. Easy-to-remove stickers are available. Why not use them, instead of alienating customers?

Had the cable-TV companies treated their customers politely like human beings (instead of crap) over the years, offering them affordable packages of services (based on need and budget), they would have engendered tremendous loyalty. Instead, when the opportunity arose, resentful customers, with long, painful memories, began cutting their cords.

3M: Innovation Archetype

3M is often cited as the prototypically innovative company. It has a long, established culture of innovation.

Jamie Northup was an engineering manager at DISTek Integration in June 2018, when he attended, as a member of the Association of Equipment Manufacturers, "Building a Culture of Innovation" at 3M's Innovation Center in St. Paul, Minnesota. Here are 3M's five keys to innovation success, according to Northup:

1. **Feel your customer's pain.** 3M considers closeness to customers a main competitive advantage with 75 customer innovation centers around the world. 3M engineers understand customer product needs by getting out to the customer *in their natural environment and spending time with them.* They look for and listen to their customers' "pain points." 3M insists that customers must inspire your innovation, and when you listen to them, customers will actually help you solve their problems.

2. **Empower employees.** Allow employees to work on what they are passionate about. Give your employees license to experiment and work on projects that they love. Promote innovative ideas and empower employees to look at their daily tasks through an innovation lens. Employees should be acknowledged and rewarded for innovative ideas.

3. **Dedicate time to innovation**. In 1948, 3M launched its 15-percent program, where 15 percent of employees' time was dedicated to innovation. The Post-It note was invented during 15 percent time. Organizations such as Hewlett-Packard and Google have both replicated this approach. Gmail and Google Earth were conceived during Google's 20-percent time. While people are encouraged to work on their ideas, they must complete their billable work first. When innovative ideas make it through the ideation phase, employees are given technology grants to pursue these ideas.

4. **Collaborative Platforms**. Use newer technologies [Slack, Microsoft Teams] to promote collaboration across the enterprise. Crowdsource good ideas within the organization and capitalize on those innovations. Establish committees to grant monies to individuals and groups for their new ideas.

5. **Attract good talent.** Showcase and highlight the innovation programs that your organization has implemented. Present videos, papers, or demos of the technologies you offer and pursue.

3M clearly demonstrates that branding, intrabranding, and innovation are inextricably linked.

CHAPTER TEN

Securing Your Keystone

A trending *Fortune* article on July 30, 2020: "Apple and Samsung Have Topped the Smartphone Market for 9 Years. Now There's a New Leader [Huawei]."

It's neverending, the misuse of terminology. **There is** *no* **smartphone market**. A smartphone *industry*, yes. Market, no. Words matter. Business media perpetuate the ignorance problem that impedes successful intrabranding, the keystone of corporate agility.

Fortune, while slobbering all over Huawei's climb to beat Samsung in smartphone shipments, *never* mentioned its direct tie to the Chinese Communist Party (CCP). The CCP spies on people all over the world via electronic backdoors in all of Huawei's products. Kind of a big deal, right?

Agility Requires Reality

By burying one's head in the sand and pretending the truth doesn't exist, how can one educate readers, let alone lead a company?

Agility requires reality. Agility's roots are urgency, alignment, and communication. Can a company become agile by ignoring reality? It can't. Corporate roadkill is everywhere.

When employees start deceiving and dodging each other, keeping secrets (like the Boeing engineers), and living in an alternative universe, behavior that CEOs cause and tolerate, enterprises become *entroprises.*

I asked a top exec of a major industrial company, reporting directly to the CEO, why the multiple divisions of his company didn't unify behind a single brand. He told me that said divisions tend to do their own things and don't like taking direction from HQ. Incredulous, I retorted that the CEO's job, like that of an orchestra conductor, is to keep all employees playing the same music, whether they like it or not.

In fairness to the aforementioned company, no employee would have been able to conform to the brand, because it didn't exist. Yet, everyone in said company knows the LGBTQ policy, the social-distancing policy, the mask-wearing policy, and likely supports the Marxist BLM.

Why is this? Wokeness is narcissistic, about self-congratulation. Branding, conversely, is inconveniently about someone else, the customer.

Failure: Wokeness, yes. Brand, no. Becoming woke at warp-speed. Branding at a snail's pace.

As stated in the Introduction, intrabranding—selling and enforcing the brand internally—is the key to successful branding, and effective communication is the foundation of intrabranding.

This book is premised on the arch-and-keystone configuration, an illustrative metaphor for corporate strength. Your enterprise, the arch, is only as strong as its keystone, intrabranding. Deny, ignore, and mock intrabranding at your peril and your shareholders' peril.

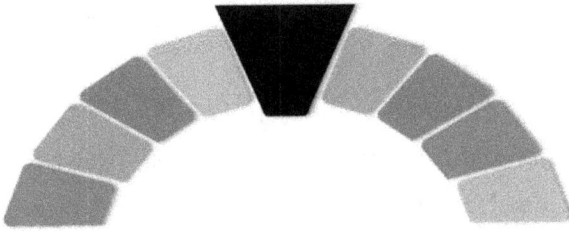

Wobbly Trader Joe's

Too many CEOs seem to fear being labeled as racists or conservatives more than being tagged as incompetents. I urge to you emulate Bob Unanue, CEO of Goya Foods, who stood up to the mob (see Chapter 3), not Dan Bane, the wobbly CEO of Trader Joe's.

Recently, a woke highschool senior, Briones Bedell, organized a Change.org petition to force Trader Joe's to stop its decades-long use of Trader Giotto's, Trader José's, and Trader Ming's on international foods.

Why? Like most kids these days, Ms. Bedell has no life other than being perpetually outraged and offended, and demanding conformity to leftist orthodoxy.

She tweeted to Trader Joe's: "The carefully-crafted facade of your friendly neighborhood hipster grocery store belies a darker image; one that romanticizes imperialism, fetishizes native cultures, and casually misappropriates."

At first, **because of one protester** (remember Niel Golightly's demise at Boeing? See Chapter 8), Dan Bane caved. That's what today's CEOs do.

Then, customers—*who, unlike the outraged petitioner, weren't offended by these "foreign" labels*—weighed in to stop the nonsense. Trader Joe's reversed itself and kept the names: "Recently, we have heard from many customers reaffirming that these name variations are largely viewed in exactly the way they were intended-—as an attempt to have fun with our product marketing."

Why didn't Trader Joe's already know its customers' views? Why did it cave into the mob, then reverse itself? The answers are easy and obvious, if you've read this entire book.

Parting Advice

You can't lead your company unless *you* are in charge and can communicate—and the mobs, internal and external, respect you. But, if the mobs are in control, your keystone is fragile—as is your arch (and your tenure). In that case, reread this book, while you still have time.

ABOUT THE AUTHOR

Marc Rudov is a branding advisor to CEOs, media commentator, and author of *Brand Is Destiny: The Ultimate Bottom Line, Be Unique or Be Ignored: The CEO's Guide to Branding*, and numerous articles. Rudov has headed marketing orgs in both large and small companies.

Known worldwide as an independent thinker and thought-leader, he is unfazed by political correctness and technological correctness.

Mr. Rudov rails against industry, product, and technology jargon, and urges his clients—from various industries—to escape their comfort zones to stand out, to be unique.

He counsels CEOs that, if they fail to lead and enforce their branding initiatives internally—the essence of intrabranding—they will create *entroprises*, imperil their destinies, and, consequently, squash their bottom lines.

Mr. Rudov holds an electrical engineering degree from the University of Pittsburgh and an MBA from Boston University.

Contact him at MarcRudov.com for advisory services, books, media appearances, debates, and speaking engagements.

www.ingramcontent.com/pod-product-compliance
Lightning Source LLC
Chambersburg PA
CBHW071423210326
41597CB00020B/3633